SpringerBriefs in Archaeol

Archaeological Heritage Management

For further volumes:
http://www.springer.com/series/10186

Douglas C. Comer • Michael J. Harrower

Mapping Archaeological Landscapes from Space

 Springer

Douglas C. Comer
International Committee on Archaeological
Heritage Management (ICAHM)
Baltimore, MD, USA

Michael J. Harrower
Johns Hopkins University
Near Eastern Studies
Baltimore, MD, USA

ISSN 1861-6623   ISSN 2192-4910 (electronic)
ISBN 978-1-4614-6073-2 ISBN 978-1-4614-6074-9 (eBook)
DOI 10.1007/978-1-4614-6074-9
Springer New York Heidelberg Dordrecht London

Library of Congress Control Number: 2012951335

Springer is part of Springer Science+Business Media (www.springer.com)

Published in Observance of the 40th Anniversary of the World Heritage Convention

Foreword

Archaeologists have made use of the vertical perspective and imaging technologies since the 1880s when dry plates made photography from hot air balloons more feasible. The use of aircraft in the 1900s enabled more widespread local to regional scale coverage for aerial photography. The last half of the twentieth century brought transformative changes with access to space, technologies to image much more of the spectrum than seen by the human eye, precise global positioning, and well-calibrated digital imaging approaches that collectively provide scientific quality spectral and spatial information across vast areas, on a repeat basis and with fine resolution. The application of these transformative technologies to archaeological discovery, to understanding the spatial relationships of archaeological finds to each other and to the physical environment, and to the preservation of cultural patrimony defines Space Archaeology.

The primary objectives of NASA's Space Archaeology Program have been to facilitate enhanced access to and use of remote sensing data, to promote the development and application of analysis tools and techniques for archaeological research, and to encourage more widespread utilization of such data and techniques. NASA data began being used for archaeological studies during the 1970s, with data streams from the Earth Resources Technology Satellites (ERTS), now the Landsat series, and the L-band SeaSAT synthetic aperture radar (SAR). The NASA Shuttle Imaging Radar(s), SIR-A/B/C, specifically targeted sites of archaeological interest. Today, Hyperion and the Advanced Land Imager (ALI) continue to provide data to the archaeological community. In addition, NASA has flown airborne systems such as SAR and LiDAR to support archaeological research in Southeast Asia and Latin America. These data supplement other space-based data streams from both the commercial sector, including both high-resolution optical data and SAR data, and more moderate resolution optical and SAR data provided by other space agencies around the world. Never before has the archaeological community been presented with such an array of sensing techniques, large archives of data global in scope with much of it publicly available, and associated visualization and analysis tools.

NASA has supported archaeological investigations since the 1980s, and these are now solicited regularly through a formal program. The inspiration for the NASA Space Archaeology Program came from the successful use of imagery obtained by NASA satellites and aerial platforms over the past four decades. Over the years, these results have been widely disseminated through professional meetings, the journal literature, a series of specialty workshops, and books. The first conference on Remote Sensing in Archaeology was held at the NASA Stennis Flight Center in 1984. The most recent NASA Space Archaeology Workshop was hosted by the Johns Hopkins University Department of Near Eastern Studies from October 8 to 10, 2011, and co-sponsored by ICOMOS International Committee on Archaeological Heritage Management (ICAHM) and the CSRM Foundation. This book was conceived at that workshop.

Much of what you will read in this volume has been developed under NASA Space Archaeology Program grants or by participants in the Space Archaeology Workshops. The book provides sections on the history of space archaeology, multispectral and hyperspectral data, synthetic aperture radar, LiDAR, and modeling. This book is intended to provide a clear and concise introduction to archaeologists who are considering the use of satellite and aerial imagery and the technologies that produce it. The authors give exciting examples of how these technologies have benefitted archaeological research and furthermore have advanced the effective management of archaeological sites and landscapes around the world. NASA looks forward to continuing collaboration with the archaeological community as more archaeologists become familiar with Space Archaeology.

Washington, DC, USA   Myron Craig Dobson

Acknowledgements

We would like to thank a number of people and institutions that played extremely important roles in the production of this publication. Prof. Glenn Schwartz of the Department of Near Eastern Studies at The Johns Hopkins University suggested that the co-editors for this book meet to discuss their mutual interest in aerial and satellite remote sending some years ago. It was through these conversations that the NASA Space Archaeology workshop was held at Johns Hopkins University October 8–10, 2011. Dr. Myron Craig Dobson, Space Archaeology Division Program Manager, Earth Science Division, at NASA Headquarters in Washington, D.C. arranged for the participation of several scientists and engineers with the NASA Jet Propulsion Laboratory at Caltech (JPL/NASA), who also wrote introductory sections on key technologies for this book. We thank Dr. Dobson for his unceasing efforts to introduce the archaeological community to technologies and imagery sources developed by NASA programs.

Several interns for the ICOMOS International Scientific Committee on Archaeological Heritage Management (ICAHM) participated in the daunting task of tracking and editing the multiple drafts of each of the 22 chapters that appear in this book. Victoria van der Haas (Leiden University) oversaw the tracking of drafts and reviews, and Robert Goldberg (UCLA) provided editing assistance, including review and revision of standard formatting throughout the book. Melanie Kingsley (Brandeis University) provided assistance to these efforts. A number of Johns Hopkins University graduate students assisted with the arrangements for the October 2011 workshop, including Jennifer Swerida, Karen (Maggie) Bryson, Sarah Yukich, and Ioana Dumitru.

We would also like to thank Prof. Fred Limp, University of Arkansas, Center for Advanced Spatial Technologies, for his comments on the manuscripts of this publication, and the authors of the chapters in this book, who each provided comments on several chapters written by others.

Contents

Contributors

Michael J. Abrams The NASA Jet Propulsion Laboratory at Caltech (JPL/NASA)

Devin Alan White Oak Ridge National Laboratory, Oak Ridge, TN, and Crow Canyon Archaeological Center, Cortez, CO, USA

Ronald G. Blom Jet Propulsion Laboratory, California Institute of Technology, La Cañada Flintridge, CA, USA

Scott Branting University of Chicago, Chicago, IL, USA

Jesse Casana University of Arkansas, Fayetteville, AR, USA

Bruce Chapman Jet Propulsion Laboratory, California Institute of Technology, La Cañada Flintridge, CA, USA

Arlen F. Chase University of Central Florida, Orlando, FL, USA

Diane Z. Chase University of Central Florida, Orlando, FL, USA

Li Chen Department of Applied Mathematics and Statistics, Johns Hopkins University, Baltimore, MD, USA

Douglas C. Comer Cultural Site Research and Management, Baltimore, MD, USA

Jackson Cothren University of Arkansas, Fayetteville, AR, USA

Li Min Cotsen Institute of Archaeology UCLA, Los Angeles, CA, USA

Bryce Davenport Department of Anthropology, Brandeis University, Waltham, MA, USA

Stephen Davis University College Dublin, Co. Dublin, Ireland

Christopher T. Fisher Anthropology, Colorado State University, Fort Collins, CO, USA

Prem Goel The Ohio State University, Columbus, OH, USA

Charles Golden Department of Anthropology, Brandeis University, Waltham, MA, USA

Michael J. Harrower Department of Near Eastern Studies, Johns Hopkins University, Baltimore, MD, USA

David R. Hixson Shepherd University, Shepherdstown, WV, USA

Ian W. Jones UC San Diego, La Jolla, CA, USA

Stephen J. Leisz Department of Anthropology, Colorado State University, Fort Collins, CO, USA

Thomas E. Levy UC San Diego, La Jolla, CA, USA

Joy McCorriston The Ohio State University, Columbus, OH, USA

William Megarry University College Dublin, Co. Dublin, Ireland

Bjoern H. Menze ETH Zurich, Zurich, Switzerland

Carey E. Priebe Department of Applied Mathematics and Statistics, Johns Hopkins University, Baltimore, MD 21218, USA

Stephen H. Savage Arizona State University, Tempe, AZ, USA

Jared Schuetter Battelle Memorial Institute, Columbus, OH, USA

Matthew Senn The Ohio State University, Columbus, OH, USA

Daniel Sussman Department of Applied Mathematics and Statistics, Johns Hopkins University, Baltimore, MD 21218, USA

James C. Tilton NASA Goddard Space Flight Center, Greenbelt, MD, USA

Jason A. Ur Harvard University, Cambridge, MA, USA

John F. Weishampel Orlando, FL, USA

Chapter 1
Introduction: The History and Future of Geospatial and Space Technologies in Archaeology

Michael J. Harrower and Douglas C. Comer

In a contemporary era transformed by computers and the Internet, geospatial sciences and space technologies will undoubtedly continue to play an increasing role in understanding and preserving archaeological histories. Particularly since the early 1990s, advances in air and spaceborne technologies, image availability, hardware and software have contributed new, substantially more effective means of archaeological research and heritage management. Geospatial technologies have begun to transform many fields, from engineering and environmental studies to health and earth sciences, and they are having similarly broad impacts in archaeology. Applications encouraged by the availability of inexpensive Global Positioning System (GPS) receivers, satellite imagery, and Geographic Information Systems (GIS) software, have reached near ubiquity in archaeological field research, yet there are few advanced introductions to space technologies tailored to the needs of archaeologists.

This volume examines the contemporary range of available air and spaceborne imagery and associated applications arranged in five sections: (1) historic air and spaceborne imagery, (2) multispectral and hyperspectral imagery, (3) SAR (Synthetic Aperature Radar), (4) LiDAR (Light Detection and Ranging), and (5) associated archaeological site detection and modeling. Each of the aforementioned sections is introduced by expert scientists who concisely describe fundamental concepts required to apply particularly datasets, followed by two or more exemplary archaeological applications in different contexts worldwide. Targeted to the needs of leading researchers and scholars in archaeology as well as graduate and advanced undergraduate students, we aim to convey a sense of what is currently possible and, it is hoped, to inspire new pioneering applications.

All archaeologists are familiar with the need to establish the provenience and context of artifacts, features, structures, and the wide variety of organic remains encountered during archaeological investigations. In rigorously establishing provenience, context, and location, geospatial technologies are increasingly essential to archaeology; without a basis for documentation of context archaeological fieldwork would be merely an exercise in the collection of old objects able to convey only a very limited understanding of the human past.

D.C. Comer and M.J. Harrower, *Mapping Archaeological Landscapes from Space*,
SpringerBriefs in Archaeology, DOI 10.1007/978-1-4614-6074-9_1,
© Springer Science+Business Media New York 2013

As archaeology has matured, the benefit of seeing archaeological materials and sites in the wider context of physical and human geographies not limited to individual sites themselves has become clear. As focus has shifted to involve landscapes, elements of locational context remain central and are substantially better established and analyzed through applications of space technologies.

In particular, archaeological survey, regional analysis, landscape archaeology and archaeological heritage management are substantially enriched through geospatial and space technologies. In the Near East, regional surveys like those pioneered in the 1930s by Nelson Glueck (1951) in the Levant, Robert Braidwood (1937) across the Amuq Plain (Turkey) (Braidwood 1937) and Thorklid Jacobsen (1958) in Iraq were among the early exploratory work that was pivotal in establishing ancient geographies. In the 1950s and 1960s, Willey's (1953) regional investigations of the Viru Valley (Peru) and Adams' (1965) air-photo assisted surveys of Iraq were similarly seminal in establishing new methods for exploring and analyzing settlement patterns. Lewis Binford famously argued (1964: 425) that "... the methodology most appropriate for the task of studying processes of cultural change and evolution is one which is regional in scope ...." Although processual archaeology has since, to some, fallen from favor, his timing for both research and heritage management was apposite: In the United States, with passage of the National Historic Preservation Act of 1966 (as amended), it became inevitable that broad surveys would become the rule rather than the exception. The Act laid the groundwork for inventories of archaeological sites over wide areas, followed by an evaluation of the significance of discovered sites, and finally establishing a list or gazette of the most important sites. This inventory and evaluation approach is used today around the world. It is, in fact, explicitly required of countries that sign *The Convention Concerning the Protection of the World Cultural and Natural Heritage*. Article 1, Section 1, of the Convention states:

> Every State Party to this Convention shall, in so far as possible, submit to the World Heritage Committee an inventory of property forming part of the cultural and natural heritage, situated in its territory and suitable for inclusion in the list provided for in paragraph 2 of this Article [the World Heritage List]. This inventory, which shall not be considered exhaustive, shall include documentation about the location of the property in question and its significance.

Regional, as opposed to solely intra-site, context thus became evermore pertinent to both archaeological research and heritage management. As Dunnell and Dancey noted (1983: 267), "...both systems theory and ecological approaches require distribution data that can be correlated in the dimension of space with other cultural, biotic, and physical variables." Similarly, the more recent birth of landscape archaeology, including interest in how humans inhabited landscapes and generated socially inscribed notions of routes and places (e.g. Tilley 1994) is similarly well informed by geospatial technologies (e.g. Llobera 2011; Lock 2010).

Clearly, the broadest possible landscape context is of great value in understanding archaeological histories and geospatially informed representations of regional and landscape contexts broadens opportunities for understandings and interpretations of all sorts. In the realm of archaeological heritage management, landscape context is important for the same essential reason: locational context helps managers apprehend the human geography of sites and the degree to which they are special, unusual, or unique. To be inscribed on a gazette of the most significant sites, whether this is the National Register of Historic Places in the United States or the World Heritage List, requires comparison of a site with others. As stated, for example, in the *Operational Guidelines for the Implementation of the World Heritage Convention* (132, 3):

> A comparative analysis of the property in relation to similar properties, whether or not on the World Heritage List, both at the national and international levels, shall also be provided. The comparative analysis shall explain the importance of the nominated property in its national and international context.

Methodologically, geospatial and space technologies have substantially enhanced archaeological investigations of a great variety of diverse aims by providing: (1) advanced means of site prospection/detection in air and spaceborne imagery, (2) improved capacity to characterize environments particularly with multispectral and hyperspectral sensors, (3) efficient Global Navigation Satellite Systems (GNSS) tools for field mapping, most recognizably GPS, but also more recently Russian GLONASS and European Galileo systems, and (4) GIS tools for advanced analysis, visualization, management and modeling of archaeological data. In the five sections that follow, we briefly review the history of air and spaceborne imagery, multispectral and hyperspectral imagery, synthetic aperture radar, LiDAR, and archaeological site detection and modeling highlighting key concepts and applicants of these technologies.

1.1 Historic Air and Spaceborne Imagery

In Section 1 of this book we begin by examining the use of imagery acquired by air- and space-borne platforms that are now historic. Stephen Leisz in Chapter 2 provides an overview – an introduction to some and a refresher to others – of the means by which views of landscapes were obtained over the past century. Only optical technology was available at the beginning of the twentieth century, elevated first by balloons and aircraft and later by satellites. By the end of the century, spectral, synthetic aperture radar (SAR), and LiDAR (light detection and ranging) sensing apparatuses were in use. In Chapter 3, Ur outlines how enormously informative "historic" optical images are when analyzed with the assistance of contemporary computer technologies. Declassified CORONA satellite imagery can be obtained at extremely low cost. Ur explains how he used such imagery to develop an inventory of hundreds of archaeological sites, over 6,000 km of pre-modern track ways, and many hundreds of kilometers of irrigation canals.

In Chapter 4, the Near East is also the area studied by Jesse Casana and Jackson Cothren by means of CORONA satellite imagery obtained during the Cold War, from 1960 to 1972. Through advanced camera model development, orthorectification computerized stereo analysis they have successfully identified many sites and have smoothed much of the difficult groundwork for future researchers. The efforts of Ur, Casana, and Cothren are of extraordinary value in ways that could not have been predicted before the outbreak of unrest in many places in the Near East, which has unfortunately been accompanied in many cases by looting. In additional to their enormous research importance, their inventories provide means to help identify, evaluate, and hopefully in some cases prevent damage caused by destruction and looting. In Chap. 5, Min describes applications of CORONA in China, which provides an invaluable research and heritage management tool, particularly as archaeological resources are increasingly threatened by the dramatic recent pace of economic development in the region.

1.2 Multispectral and Hyperspectral Imagery

Michael Abrams of the NASA Jet Propulsion Laboratory at Caltech (JPL/NASA) and Douglas C. Comer introduce Section 2 of this book with a concise overview of the technologies used to collect and analyze multi- and hyperspectral (reflected visible, infrared, and thermal infrared) imagery. Depending on the desired application, it is often necessary to radiometrically, spectrally or spatially correct or enhance data to make files sizes smaller, prepare for further analysis, or simply to make images visually more appealing and comprehensible to viewers. Abrams offers an informative brief description of how sensors operate along with some of the ways image interpretation has proven most useful to archaeologists with most attention given to low end-user cost sources. Douglas C. Comer provides an example of how free Landsat imagery was used to identify regions of arable soils around Petra in Chap. 7. He argues that the shift from nomadism to a more settled monument building society coincides with great cultural changes during the first and second centuries A.D. including the adoption of agriculture, which can be greatly informed through space technologies.

Megarry and Davis apply LiDAR and Worldview-2 imagery to study the vicinity of another World Heritage Site, 'Bru na Bóinne (Ireland), in Chap. 8 demonstrating the enormous utility high resolution sensors even in landscapes that have already been studied intensively. In a pioneering application of hyperspectral imagery, Stephen Savage (Chap. 9) uses NASA's Hyperion imagery to identify ancient copper mining and processing in Jordan – an enormously important activity that began in the area during the third millennium B.C. and transformed surrounding landscapes.

1.3 SAR (Synthetic Aperture Radar)

Section 3 of this book examines synthetic aperture radar (SAR), which offers a versatile toolkit that can greatly assist in detecting a wide range of natural landscape and archaeological features. Radar can image subjects as varied as topography, vegetative structure, and even superficial roughness in studies of ice or geologic formations. If SAR data are collected and analyzed interferometrically, they can also be used to develop digital surface models of the ground or vegetative canopies, depending on wavelength. SAR returns are influenced by the capacity of materials encountered to conduct electricity, are generally unaffected by cloud cover, and under certain conditions can penetrate vegetation and soil to reveal what lies beneath. In Chap. 10, which begins this section, Bruce D. Chapman and Ronald G. Blom of JPL/NASA describe these and other characteristics of SAR, and provide examples of the ways in which SAR's unique capacities have and can be used for archaeology.

In Chap. 11, the use of SAR to detect archaeological features under vegetative cover in the Mayan wetness is described and discussed by David Hixson. Hixson used SAR data collected in a 2004 deployment in Central America of the JPL/NASA aerial platform AirSAR, which carried an apparatus to transmit and receive SAR at long and short wavelengths (approximate wavelengths: P-Band, 75 cm, L-Band, 25 cm, C-Band, 5 cm). Each of these could be polarized either vertically or horizontally at transmission or reception. This multi-polar capacity enhances capability to detect different aspects of target structure; in this case, large mounded architecture. Hixson also used Landsat imagery, which was also successful in detecting site locations, and he describes the strengths and weaknesses of both types of imagery in densely vegetated environments. Charles Golden and Bryce Davenport in Chap. 12 similarly evaluate the utility of AirSAR data collected during the 2004 deployment in their research area along the Usumacinta River, which separates Mexico from Guatemala, from Yaxchillan in the south to Piedras Negras in the north. They found the digital surface models generated by interferometric analysis of AirSAR data to be more useful in modeling viewsheds when compared with other readily available Digital Elevation Models (DEMs) including those produced from ASTER and SRTM.

In Chap. 13, Douglas C. Comer and Ronald G. Blom describe successfully detecting numerous archaeological sites on San Clemente Island, one of the Southern Channel Islands off the coast of Southern California with the use of AirSAR. They discuss the cultural implications of the detected site distribution pattern, which is strongly linked to viewsheds interferometrically produced from SAR. Viewsheds or viewscapes appear to have been a dominane factor in the selection of habitation sites as they allowed surveillance of the surrounding ocean to identify resources, especially sea mammals that were only available for short periods and involved coordination of groups to harvest them.

1.4   LiDAR (Light Detection and Ranging)

Devin White provides an overview LiDAR technology tailored to archaeological research and heritage management to introduce Section 4 of this book. White discusses point clouds, which in some cases can be produced by technologies other than LiDAR such as digital photographs. However multi-dimensional point clouds are generated, they can be used for a wide range of products such as surface models, contour lines, viewsheds, least-cost paths, and 3D fly-throughs. One of the applications of LiDAR technology in archaeology that has attracted great attention recently is described by Arlen Chase in Chap. 15. Funded by a NASA Space Archaeology grant, Chase obtained LiDAR for a 200 km² area that included the archaeological site of Caracol, in Belize. This high-resolution LiDAR coverage, at 20 points per m², was enough to produce a "bald-earth" model of the landscape; that is, the surface that lay beneath the dense, tropical vegetation that covered the area. As Chase notes, because of the nearly impenetrable tropical vegetation surrounding the central monuments, this was the first time that an entire Mayan city could be mapped. The surface model developed from the LiDAR point clouds revealed pyramids, temples, houses, agricultural terraces, pathways and roads, ball courts, reservoirs, and in fact virtually every feature that retained surface relief in the ancient city. The applications to research are obvious but, in addition, such a model can be used to plan for the protection of the area and the development of truly sustainable tourism.

LiDAR was also used by Chris Fisher and Stephen Leisz at Purépeche, in Mexico, who identify areas used for agriculture and discuss implications for understanding the site in Chap. 16. The surface model developed from LiDAR was less precise than for Caracol because the density of LiDAR points was not as high. The dense coverage obtained for Caracol was, as one might expect, more costly. Yet the 2.5 m resolution surface model with vertical accuracy of ±2 m obtained for Purépeche informatively revealed terrace and raised field locations that covered a vast area, which can in turn be used to revise population estimates for the empire period at the site.

1.5 Archaeological Site Detection and Modeling

Michael Harrower introduces Section 5 of this book by discussing methods, concepts and challenges of archaeological site detection and modeling. He discusses the quandaries archaeologists face in defining site taxonomies, choosing between intensive (full coverage) versus extensive (regional) archaeological survey approaches and related opportunistic or systematic sampling strategies. He goes on to address a few key best practices for conducting and reporting results of predictive models and archaeological sites detection methodologies, including errors of omission (false negative), errors of commission (false positive) and The Modifiable Areal Unit Problem (MAUP). Bjoern Menze and Jason Ur in Chap. 18 demonstrate the tremendous capacity of multispectral imagery to detect ancient settlements through spectral discrimination of anthrosols. Their innovative approach and impressive results used 160 ASTER images to detect more than 14,000 sites and potential sites over a more than 20,000 km² area. The outcome of their analysis is an unparalleled dataset of settlement patterning that dramatically enlightens understanding of ancient urbanism across northern Mesopotamia and offers enormous future potential.

Scott Branting discusses potential future applications of geospatial technologies in Chap. 19, and employs his work at Kerkenes Dağ in Turkey as a case study. There, he uses high-resolution GPS micro-topographic data and results of more traditional excavation to model and analyze the movements of ancient people across the site. His analyses facilitate mapping of pedestrian traffic across the site, conveying a far deeper understanding of movement and use of space with vast future potential in this and other regions. James G. Tilton and Douglas C. Comer in Chap. 20 present a site detection protocol using both SAR and multispectral data (from the IKONOS satellite). This work built upon that conducted by Comer and Blom at San Clemente Island (see Chap. 13). Tilton and Comer here describe the use of the many of the same SAR and multispectral data sets, but in a very different environment. Santa Catalina Island receives twice as much rainfall as San Clemente Island (only 20 miles distant), and is an island with a much different topography of high ridges and narrow valleys. Consequently, the probability that images might be imprecisely registered was high. This necessitated an enhancement of the sampling scheme that would accommodate some degree of imprecision. A data quantization scheme that utilized image segmentation software previously developed by Tilton was incorporated in the protocols with impressive results.

In Chap. 21, Chen et al. describe research using some of the same statistical protocols developed in the work of Comer et al. (Chap. 13). Eight multispectral bands acquired by the Digital Globe's Wolrdview-2 satellite, and slope models developed from LiDAR, were used to examine Fort Irwin, California. Statistical protocols include principal components analysis, followed by a linear discriminate analysis that informs a classifier that assigns a probability that a location is or is not an archaeological site. Results obtained from a suite detection protocols were further tested against an archaeological predictive model for Fort Irwin. Harrower et al. describe high-precision GPS survey of south Arabian tombs, development of an automated algorithm to detect them in Quickbird imagery, and preliminary GIS analysis of their distributions in Chap. 22. Many thousands of such small circular stone tombs are to be found across southern Arabia (with analogous examples as far spread as the Sinai and Syria); Harrower and team emphasize that in this and other regions satellite imagery is the only practical way to comprehensively document such numerous small-scale monuments over large areas.

References

Adams, R. M. (1965). *Land behind Baghdad: A history of settlement on the Diyala Plains*. Chicago: University of Chicago.

Binford, L. R. (1964). A consideration of archaeological research design. *American Antiquity, 29,* 425–441.

Braidwood, R. J. (1937). *Mounds in the plain of Antioch: An archaeological survey*. Chicago: University of Chicago Oriental Institute.

Dunnell, R. C., & Dancey, W. S. (1983). The siteless survey: A regional scale data collection strategy. *Advances in Archaeological Method and Theory, 6*, 267–287.

Glueck, N. (1951). *Explorations in Eastern Palestine* (The annual of the American Schools of Oriental Research). New Haven: American Schools of Oriental Research.

Jacobsen, T. & R. M. Adams (1958). Salt and Silt in Ancient Mesopotamian Agriculture. *Science* 128(3334): 1251–1258.

Llobera, M. (2011). Archaeological visualization: Towards an archaeological information science (AISc). *Journal of Archaeological Method and Theory, 18*, 193–223.

Lock, G. R. (2010). Representations of space and place in the humanities. In D. J. Bodenhamer, J. Corrigan, & T. Harris (Eds.), *The spatial humanities: GIS and the future of humanities scholarship* (pp. 89–108). Bloomington: Indiana University Press.

Tilley, C. (1994). *The phenomenology of landscape*. Oxford: Berg.

Willey, G. (1953). *Prehistoric settlement patterns in the Viru Valley, Peru*. Washington, DC: US Government Printing Office.

Part I
Historic Air and Spaceborne Imagery

Chapter 2
An Overview of the Application of Remote Sensing to Archaeology During the Twentieth Century

Stephen J. Leisz

Abstract Powered flight was perfected at the dawn of the twentieth century and provided us with a bird's eye view of large swaths of landscape for the first time in human history. Over the course of the twentieth century, this bird's eye view expanded to space, as the frontier of powered flight was pushed, literally, to the moon. This chapter overviews the role that advances in aerial photography and imaging and spaceborne photography and imaging played in the advancement of archaeological research in the twentieth century. It is noted that through the twentieth century most applications of remote sensing to archaeology have involved data from one remote sensing platform at a time. As the century came to a close, and new remote sensing technologies such as LiDAR were being introduced to archaeology, some started to realize that the most promising aspect of applying these technologies to archaeology may be how they can be integrated.

Keywords Remote sensing • Twentieth century • Aerial photography • Spaceborne imagery

2.1 Introduction

Over the course of the twentieth century archaeologists have made use of many remote sensing technologies. This chapter looks at how advances in photographic and imaging technologies have been used by archaeologists in the twentieth century. The role that aerial photography played in archaeological prospecting and mapping from its inception through the 1960s is first overviewed. Next a brief overview of space based photography and imagery is presented along with the more recent use of radar.

D.C. Comer and M.J. Harrower, *Mapping Archaeological Landscapes from Space*, SpringerBriefs in Archaeology, DOI 10.1007/978-1-4614-6074-9_2, © Springer Science+Business Media New York 2013

2.2   Aerial Photographs and Archaeology, 1908 to the 1960s

From 1839, when photography was born, photographers have sought to carry cameras aloft to obtain a bird's-eye view of the earth's surface. From 1858 when Parisian
photographer Gaspard-felix Tournachon took a photograph from a tethered balloon
over Val de Bievre near Paris, to 1908 when Wilbur Wright took aerial motion pictures above Le Mans, France, photographs from the air were taken from kite, balloon, and even "bird" platforms (Lillesand et al. 2008). After 1908 most photographs
taken from above were taken from airplanes, until 1959 when photographs and,
later, images of the earth's surface were also taken from high earth orbit. In each of
these eras archaeologists have recognized the potential of gaining a "bird's-eye"
view of the landscape in order to better identify sites and remains.

The utilization of aerial photographs for military and mapping purposes drove
the development of techniques for interpreting air photographs and utilizing the
information to more quickly carry out tasks, such as reconnaissance and map creation, that had previously been accomplished through terrestrial surveying (Thomas
1945). One of the earliest published reports of aerial photography applied to archaeology came out of the intersection of using air photos to survey an area as part of
map production in support of military purposes. Lieutenant-Colonel Beazeley
(1919) reported on the use of aerial photography in Mesopotamia as part of a military land survey during the First World War. The objective of the air survey was to
quickly topographically map the region. Beazeley notes a number of archaeological
discoveries as part of the survey: the inadvertent discovery of "the remains of an
ancient city" (Samarra) and the "outline of a series of detached forts" visible from
the air, while "no trace was visible" from the ground (Beazeley 1919).

From World War I until after World War II, aerial surveying was further developed in support of archaeology in many parts of the world. In Great Britain and
Europe, O. G. S. Crawford detailed the utility of air photos to "reveal earthworks
upon ploughed land which are invisible to the observer on the ground," (Crawford
1923). Along with Crawford, Alexander Keiller and Major Allen used aerial photography in Central Wessex, England, to uncover a number of archaeological sites, such
as camps, dykes, lynchets, crop circles, and extinct agricultural villages (St. Joseph
1945; Palmer 1947). They also made use of the unique perspective of aerial photographs to analyze the relationships between these sites. In Central and South America,
the 1920s and 1930s saw the use of aerial photography to complement ground surveys. Ricketson and Kidder (1930) document the flights made by Colonel Charles
Lindbergh in October 1929 as part of the Carnegie Institution of Washington's surveys in the Maya area of British Honduras, Guatemala, and southern Mexico.
Oblique photography from the air was done and interpretation of the air photos
yielded identification of both known and previously unknown archaeological sites.
In Peru the 1931 Shippee-Johnson expedition was the "first systematic attempt to
use aerial photography to discover, locate and describe" archaeological sites in South
America (Denevan 1993). Shippee-Johnson made use of vertical and oblique aerial
photography as well as ground photography. Aerial photography was also used by

Fig. 2.1 Near vertical view of West Baray, Angkor. Courtesy of the Williams-Hunt aerial photos Collection, Digital Archive Center for Southeast Asia Studies (CSEAS), University of Kyoto

archaeologists in other geographic settings during the period between World War I and World War II, such as Central and Eastern Europe, North Africa, and Asia.

Post-World War II saw a further blossoming of the use of aerial photography for archaeological purposes. At least part of this can be attributed to the use of aerial surveillance during the war and the acquisition through these means of air photos for places that otherwise would not have been surveyed (Chapman 1945). An example is the Williams-Hunt aerial photograph collection (Moore 2009). Peter Williams-Hunt flew reconnaissance missions for the Royal Air Force in Southeast Asia over Myanmar (Burma), Cambodia, Laos, Thailand, Singapore and Vietnam. He had been interested in the use of aerial photographs for archaeological purposes (Bradford and Williams-Hunt 1946) and his collection of reconnaissance photographs has proven useful to archaeological work across Southeast Asia (Moore 2009). Figure 2.1 showing the West Baray, Angkor Wat is illustrative of this. During this time, archaeological uses of aerial photographs spread into Oceania. Palmer describes the use of aerial photography in support of archaeology in

New Zealand (1947) to indentify Maori *pa* and terraces. During the 1950s, the use of aerial photography by archaeologists also expanded in sub-Saharan Africa (Meighan et al. 1958). The use of aerial photography in documenting archaeological sites worldwide has continued to the present day (Lillesand et al. 2008).

2.3   Post 1950s Through the End of the Twentieth Century: Photography from Space, Satellite Imagery, and Radar

From the early 1960s to end of the twentieth century the platform for the "bird's eye" view of the earth moved from air to space. While photographs taken from airborne and spaceborne platforms continued to be used by archaeologists, multi-spectral imaging systems (which recorded information from non-visible wavelengths such as near, middle, and thermal infrared) and radar were introduced to archaeologists.

The first spaceborne platform was introduced in 1959 when the initial satellite under the CORONA satellite program was launched by the United States military. Its primary mission was to provide military reconnaissance and because of this, even though it was the earliest earth imaging satellite instrument, its products were not available to the public or to archaeologists until President Clinton declassified the mission's products on February 22, 1995. The declassification led to the use of CORONA data in the latter 1990s to examine landscapes and ancient cities in Turkey and the Euphrates Valley (Kennedy 1998) and in Syria (Ur 2003). The CORONA satellite program's imaging instruments were photographic cameras placed into space. The best spatial (ground) resolution was 6 ft (1.8 m) (Jensen 2000, Lasaponara and Masini 2011), while the coarsest pictures had spatial resolutions of approximately 460 ft (Jensen 2000). Once the camera had used up its film, the film was returned to Earth in a capsule via parachute and then developed. Figure 2.2 is an example of a 1967 photograph taken from a KH-4 satellite of the CORONA program. The photograph shows a section of the "Long Wall" of Central Vietnam (Hardy and Nguyễn 2011). The section of the wall is preserved in the photograph, but not visible on the ground today. This example illustrates how even though the CORONA program photographs were not available to archaeologists until decades after they were originally taken, they are still useful today.

While the CORONA historical satellite photos have been the most used by archaeologists, there are also other examples of photographs taken from space. The Russian KVR-1000 spaceborne photographs, which were made available for a limited period in the 1990s (Fowler 2004; Lasaponara and Masini 2011), were used by Fowler (1996) to detect archaeological features near Stonehenge and by Comfort (1997) to investigate the archaeological site of Zeugma in Turkey. Manned Skylab missions 1–4, which were undertaken in 1973 and 1974, have also been used by archaeologists, most notably to identify prehistoric irrigation canals in the southwestern part of the United States (Giardino 2011).

In 1972 the United States launched the Earth Resources Technology Satellite, which was later renamed Landsat. While images from this satellite were not as

Fig. 2.2 The Long Wall of Vietnam identified on a KH-4 Corona picture from 1967 of Central Vietnam. Image provided by the USGS EROS Data Center

detailed as those from the CORONA program, they were immediately available and so their use by archaeologists predated the use of CORONA. In 1984 the French SPOT program was initiated, and archaeologists also made use of SPOT image products. Initial Landsat imagery (1972–1982, Landsat 1–3) was from the "multi-spectral scanner" (MSS). This instrument collected data in the green, red, and near-infrared parts of the electromagnetic (EM) spectrum at a spatial resolution of 79 m. From 1982 onward (Landsat 4,5, and 7) the "thematic mapper" (TM) had a spatial resolution of 28.5 m and collected information in the blue, green, red, near-infrared, mid-infrared, and thermal infrared parts of the EM spectrum. SPOT collected information at higher spatial resolutions from its inception in 1984 (10 m with its pan-chromatic instrument and 20 m with its multi-spectral instrument). Its first three mission satellites collected information in the green, red and near-infrared parts of the EM spectrum; its later missions, since 1998, collect information in the green, red, near-infrared and mid-infrared parts of the EM spectrum.

Unlike aerial photography, it is not possible to identify exactly what is on the ground using medium resolution Landsat and SPOT imagery. Rather, inferences have to be made from the EM reflections and from the spatial relationships of structures that can be identified. Blom et al. (1997) did exactly this in using Landsat and SPOT data in conjunction with other remote sensing data and ancillary data to identify the "Lost City of Ubar". Others have also made use of Landsat data to identify archaeological sites. Showalter (1993) experimented with Landsat TM imagery and discovered that it proved a valuable addition to the other techniques in identifying prehistoric Hohokam canals in the American southwest. Montufo (1997) found that while Landsat TM imagery did not clearly depict rural land-use patterns on the island of Mallorca, through filtering of the digital image, linear features could be identified and the filtered product could be used to carry out landscape level analysis and detect evidence of prehistoric land-use. And El-Baz (1997), in an overview of remote sensing applied to archaeology highlighted how SPOT imagery has been used as an input to landscape level analysis for archaeological purposes in north-western Greece.

Custer (1986) takes the use of medium resolution satellite image data one step further and develops techniques for utilizing information derived from Landsat TM data in predictive models that are used to identify archaeological sites. Today the use of Landsat and SPOT imagery as inputs to predictive archaeological site models has become widespread. These examples provide an overview of the uses of medium resolution data in archaeology from the availability of these data to the end of the century. A further aspect of Landsat and SPOT data is that given the continuous availability, reliability, and affordability of the data from these sensors over a multi-decadal period, information derived from the analysis of this imagery has proven very useful in monitoring change over time at archaeological sites.

Coincident with Landsat and SPOT imagery coming online and being used by archaeologists, radar systems were being put into space, and archaeologists were experimenting with the usefulness of their products. Radar instruments transmit microwaves toward the surface of the earth; they are unaffected by clouds, and some wavelengths can penetrate vegetation and tree cover, even the top layers of soil, and then are reflected back toward the originating instrument where they are recorded on a continuum of brightness (El-Baz 1997). The "brightest" reflections represent a strong signal return and are usually from irregular, e.g. rocky, terrain; the "darkest" reflections represent a weak signal return and are usually from smooth surfaces (El-Baz 1997).

Some of the earliest uses of spaceborne radar by archaeologists were to make use of the shuttle imaging radar, SIR-A, in 1981 to identify a series of channel-like drain-ages underneath sand in the Eastern-Sahara (McCauley et al. 1982; Wendorf et al. 1987). Surveys carried out around these channels uncovered hundreds of archaeologi-cal sites (Wendorf et al. 1987). A later iteration of shuttle imaging radar, SIR-B, com-plemented information from the SIR-A radar and was used by McCauley et al. (1986) to identify what had been an area of missing drainages in the Eastern-Sahara. This led to the excavation of previously unknown archaeological sites, and in conjunction with data from a still later shuttle imaging radar mission (SIR-C), led to a better under-standing of the ancient history of the region (El-Baz 1997). Since the initial use of radar from the space shuttle platform, satellite borne radar, such as Canada's RADARSAT, has been introduced and its products have been used by archaeologists (see for example the use of RADARSAT in Mongolia by Holcomb (2001)).

One drawback to the use of radar to identify archaeological sites and surface remains is the coarseness of the spatial resolution of the radar products. As with

Landsat and SPOT imagery, the resolution is often in the tens of meters. This means that landscape level analysis can be done, but there are limits as to what individual structures can be identified. This issue was focused on in conjunction with the use of airborne radar in the Mayan region in the late 1970s and early 1980s. Adams et al. (1981) reported that they had used data from an airborne synthetic aperture radar (SAR) system to identify intensive cultivation patterns and irrigation networks beneath tree cover in the Tikal zone of Guatemala. However, Pope and Dahlin (1989) disagreed with this analysis. Pope and Dahlin's analysis of the same data used by Adams et al. concluded that "the noise in and resolution of the SAR imagery" makes it impossible to detect the cultivation pattern and irrigation network that Adams et al. report identifying. Pope and Dahlin (1989) note however that the data can be used for landscape level analysis and to identify small-scale patterns below tree-cover. The issues of overcoming noise and coarse resolutions highlighted by Pope and Dahlin with regard to the application of airborne and spaceborne radar data to archaeology is still an area of active research.

2.4   Conclusion

The twentieth century started with humans entering the age of powered flight. With powered flight came the ability to gain a bird's-eye view of the landscape and to record it. Archaeologists quickly saw the possibilities of this powerful tool and made full use of it. As the twentieth century passed its mid-point, humans sent satellites into space and used these platforms to observe the earth's surface. Archaeologists recognized the usefulness of the data produced to identify archaeological sites and remains and expanded on this functionality by incorporating the wider landscape level field of view available from spaceborne multispectral scanners such as those on Landsat and SPOT, to identify landscape level archeological features and develop predictive models of archaeological sites. As the twentieth century came to a close, high-resolution civilian multispectral satellites were being proposed, and the first, IKONOS, had been launched, carrying with it the promise of combining air photograph quality images with landscape level field of views and applying these products to archaeological research.

However, both aerial photography and satellite imagery cannot be used to identify sites obscured by tree cover or covered in sand. At the end of the twentieth century, many thought radar offered the possibility for archaeologists to take the use of remote sensing to the next level and carry out surveys from air and space that could identify remains previously hidden from them. As the twenty-first century started, another remote sensing technology, LiDAR (Light Detection and Ranging), which was only in its infancy in the late 1990s, appears to be poised to lead the next wave of remote sensing innovation within archaeology.

As this chapter is only an overview and is focusing on the progression of remote sensing's use within the field of archaeology in the twentieth century, high resolution imagery from space and LiDAR will not be further addressed. Instead the reader is referred to the chapters in this book which discuss recent applications of these newer remote sensing technologies to archaeology. If the reader would like to delve deeper into the varied applications of remote sensing and archaeology it is suggested that they consult the more thorough overviews of this subject by Parcak (2009) or Wiseman and El-Baz (2007) or the special issue of the Journal of Archaeological Science on this subject edited by Lasaponara and Masini (2011).

In closing, a final point needs to be made regarding the use of remote sensing within the field of archaeology that will remain true no matter how many new tools are added: The most promising aspect of all of these technologies may be how they can be integrated. As the price of the data products from air and spaceborne instruments are decreasing, some projects are making use of multiple sources of imagery to better understand the site being studied (see Evans et al. 2007). The interpretive power unleashed by analyzing these multiple sources of information in tandem is truly exciting.

References

Adams, R. E. W., Brown, W. E., & Culbert, T. P. (1981). Radar mapping, archeology, and ancient Maya land use. *Science, 213*(4515), 1457–1463.
Beazeley, G. A. (1919). Air photography in archaeology. *The Geographical Journal, 53*(5), 330–335.
Blom, R., Clapp, N., Zarins, J., & Hedges, G. R. (1997). Space technology and the discovery of the Lost City of Ubar. *IEEE Aerospace Conference Proceedings, 1*, 19–28.
Bradford, J., & Williams-Hunt, P. (1946). Siticulosa Apulia. *Antiquity, 20*, 191–200.
Chapman, V. J. (1945). Air photography in the Far East. *Nature, 156*(409), 604.
Comfort, A. (1997). Satellite remote sensing and archaeological survey on the Euphrates. *Archaeological Computing Newsletter, 48*, 1–8.
Crawford, O. G. S. (1923). Air survey in archaeology. *The Geographical Journal, 61*(5), 342–360.
Custer, J. F. (1986). Applications of Landsat data and synoptic remote sensing to predictive models for prehistoric archaeological sites: An example from the Delaware coastal plain. *American Antiquity, 51*(3), 572–588.
Denevan, W. M. (1993). The 1931 Shippee-Johnson aerial photography expedition to Peru. *Geographical Review, 83*(3), 238–251.
El-Baz, F. (1997). Space age archaeology. *Scientific American, 277*, 60–65.
Evans, D., Pottier, C., Fletcher, R., Hensely, S., Tapley, I., Milnes, A., & Barbetti, M. (2007). A comprehensive archaeological map of the world's largest preindustrial settlement complex at Angkor, Cambodia. *PNAS, 104*(36), 14277–14282.
Fowler, M. J. F. (1996). High-resolution satellite imagery in archaeological application: A Russian satellite photograph of the Stonehenge region. *Antiquity, 70*(1996), 667–671.
Fowler, M. J. F. (2004). Archaeology through the keyhole: The serendipity effect of aerial reconnaissance revisited. *Interdisciplinary Science Reviews, 29*(2), 118–134.
Giardino, M. J. (2011). A history of NASA remote sensing contributions to archaeology. *Journal of Archaeological Science, 38*(2011), 2003–2009.
Hardy, A., & Nguyễn, T. Đ. (2011). *Khảo cổ học Trường lũy: 5 năm nghiên cứu [The archaeology of the Long Wall: 5 years of research]* in *Những phát hiện mới về khảo cổ học năm 2010 [New archaeological discoveries in 2010]* (pp. 17–19). Hanoi: Nxb Khoa học Xã hội.

Holcomb, D. (2001). Imaging radar and archaeological survey: An example from the Gobi Desert of Southern Mongolia. *Journal of Field Archaeology, 28*(1–2), 131–141.

Jensen, J. R. (2000). *Remote sensing of the environment: An Earth resource perspective.* New Jersey: Prentice Hall.

St Joseph, J. K. (1945). Air photography and archaeology. *The Geographical Journal, 105*(1/2), 47–59.

Kennedy, D. (1998). Declassified satellite photographs and archaeology in the Middle East: Case studies from Turkey. *Antiquity, 72*(277), 553–561.

Lasaponara, R., & Masini, N. (2011). Satellite remote sensing in archaeology: Past, present and future perspectives. *Journal of Archaeological Science, 38*, 1995–2002.

Lillesand, T. M., Kiefer, R., & Chipman, J. W. (2008). *Remote sensing and interpretation* (6th ed.). New York: Wiley.

McCauley, J. F., Schaber, G. G., Breed, C. S., Grolier, M. J., Haynes, C. V., Issawi, B., Elachi, C., & Bloom, R. (1982). Subsurface valleys and geoarchaeology of the eastern Sahara revealed by shuttle radar. *Science, 218*(4576), 1004–1020.

McCauley, J. F., Breed, C. S., Schaber, G. G., McHugh, W. P., Issawi, B., Haynes, C. V., Grolier, M. J., & El Kilani, A. (1986). Paleodrainages of the eastern Sahara – The radar rivers revisited (SIR-A/B implications for a mid-tertiary trans-African drainage system). *IEEE Transactions on Geoscience and Remote Sensing, GE24*(4), 624–648.

Meighan, C. W., Pendergast, D. M., Swartz, B. K., & Wissler, M. D. (1958). Ecological interpretation in archaeology: Part 1. *American Antiquity, 24*(1), 1–23.

Montufo, A. M. (1997). The use of satellite and digital image processing in landscape archaeology. A case study from the Island of Mallorca, Spain. *Geoarchaeology: An International Journal, 12*(1), 71–85.

Moore, E. (2009). The Williams-Hunt collection: Aerial photographs and cultural landscapes in Malaysia and Southeast Asia. *Sari: International Journal of the Malay World and Civlization, 27*(2), 265–284.

Palmer, G. B. (1947). New Zealand archaeology and air photography. *The Journal of the Polynesian Society, 56*(3), 233–241.

Parcak, S. H. (2009). *Satellite remote sensing for archaeology.* New York: Routledge.

Pope, K. O., & Dahlin, B. H. (1989). Ancient Maya wetland agriculture: New insights from ecological and remote sensing research. *Journal of Field Archaeology, 16*, 87–106.

Ricketson, O., & Kidder, A. V. (1930). An archaeological reconnaissance by air in Central America. *Geographical Review, 20*(2), 177–206.

Showalter, P. S. (1993). A thematic mapper analysis of the prehistoric Hohokam canal system, Phoenix, Arizona. *Journal of Field Archaeology, 20*(1), 77–90.

Thomas, H. H. (1945). Recent developments in air photography. *Nature, 156*(3962), 409–411.

Ur, J. (2003). CORONA satellite photography and ancient road networks: A northern Mesopotamian case study. *Antiquity, 77*(295), 102–115.

Wendorf, F., Close, A. E., & Schild, R. (1987). A survey of the Egyptian radar channels: An example of applied archaeology. *Journal of Field Archaeology, 14*(1), 43–63.

Wiseman, J. R., & El-Baz, F. (2007). *Remote sensing and archaeology.* New York: Springer.

Chapter 3
CORONA Satellite Imagery and Ancient Near Eastern Landscapes

Jason A. Ur

Abstract The declassification of imagery from CORONA and subsequent intelligence satellite programs has inspired a revolution in landscape archaeology in the Near East. CORONA imagery is inexpensive, easily accessible, of high spatial resolution, and in many cases predates destructive modern development. It has become a standard tool for both remote sensing analysis and field survey. This chapter reviews how archaeologists have used CORONA to study settlement, movement, and land use via case studies from Syria, Turkey, Iran, and Iraq.

Keywords CORONA • Archaeology • Land use • Settlement • Mesopotamia • Syria • Iraq

3.1 Introduction

Archaeological remote sensing now has a century of research under its belt. After early experiments with balloon photography, aerial archaeology found its footing with the development of the airplane during the First World War. In England, aerial methods were pioneered by O. G. S. Crawford, whereas the Jesuit priest Antoine Poidebard conducted aerial reconnaissance throughout Syria (see especially Poidebard 1934). The work of Crawford, Poidebard and others revealed not only previously unknown sites but also placed them into broader cultural landscapes that included canals, roads, and relict field systems.

The easy access to aerial photography at the time of Poidebard's work is no longer possible under the post-mandate national governments in the modern Middle East. To investigate such "denied" regions, archaeologists have increasingly turned to satellite imagery. The resolution of space-based imaging systems such as Landsat and SPOT is often too coarse for archaeological features (Kouchoukos 2001). On the other hand, commercial high-resolution imagery (e.g., from the QuickBird and Ikonos satellites) is expensive and documents the modern developed landscape.

D.C. Comer and M.J. Harrower, *Mapping Archaeological Landscapes from Space*,
SpringerBriefs in Archaeology, DOI 10.1007/978-1-4614-6074-9_3,
© Springer Science+Business Media New York 2013

Since 1998, archaeologists working in the Middle East have increasingly exploited a newly available resource that is both high resolution and predates a great deal of development: declassified US intelligence imagery from the CORONA and GAMBIT programs.

CORONA acquired photographs from 1960 to 1972; its mission was to monitor Soviet missile strength, and in this it was a huge success (McDonald 1997). Features of interest identified by CORONA were further investigated by a companion "spotter" satellite, named GAMBIT (Richelson 2003). Declassified imagery from both programs can be previewed and ordered on the USGS website (http://earthexplorer. usgs.gov) or downloaded through the *CORONA Atlas of the Middle East* (http:// corona.cast.uark.edu). Archaeologists in the Near East were quick to recognize the potential of this new dataset. The first archaeological application (Kennedy 1998) examined sites in a now-inundated floodplain of the Turkish Euphrates. Systematic studies quickly followed throughout the Near East, and are reviewed below. See the contribution by Casana and Cothren (this volume) for technical details on the cameras and how they have been integrated into archaeological GIS databases.

Before reviewing several case studies from the Near East, we must consider the environment and its impact on site formation. Archaeologists have disproportionately focused on areas of early social complexity in semi-arid alluvial plains; today these areas are treeless and heavily agricultural, with a correspondingly high site visibility. Sites can be defined by three primary characteristics. The environment encourages the use of mud brick as a building material, so sites are mounded, ranging from less than a meter to more than 40 m high. Because most are currently under cultivated land, they have a high density of surface artifacts via plowing. Finally, mud brick decay and other human and domestic animal activities result in a loose soil texture that appears lighter than the background soils. Of these three characteristics, only artifact density cannot be perceived from orbit.

3.2 Settlements

In alluvial environments, CORONA imagery can be used to identify sites with high precision. Mounded sites can be discerned by their shadows (Fig. 3.1a). In the morning, their northwestern slopes are shaded, and their northeastern sides fall in shadows on images acquired in the afternoon. Southern slopes are uniformly illuminated (on modeling CORONA acquisition times, see Fowler 2006). Shadows are rarely apparent in recent commercial imagery, which is most often acquired at times that minimize shadows.

High mounds are easily recovered using traditional survey techniques. It is far more difficult to identify low- or un-mounded sites, which comprise the majority of sites. Such sites are visible on CORONA scenes because of their anthropogenic soils, which appear as lighter areas against darker natural soils (Fig. 3.1b). This phenomenon results from soil development in a parent material of decayed mud brick, hence an indicator of abandoned architecture (Wilkinson et al. 2006).

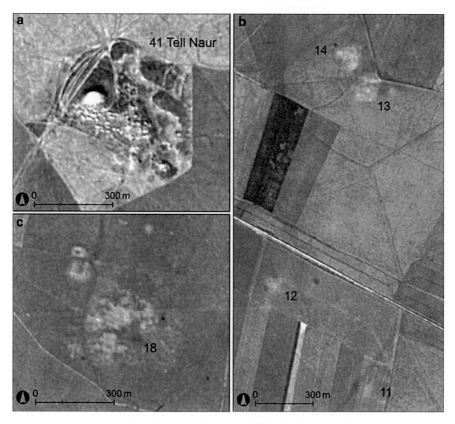

Fig. 3.1 Site signatures on CORONA imagery (1108-1025DA005-A006, 2 Dec 1969). (**a**) High mounded site; (**b**) Low mounded sites; (**c**) Complex mounded site. All sites from the Tell Hamoukar Survey, northeastern Syria (Ur 2010)

Relatively recent sites (i.e., the last two millennia) have a complex morphology of mounded areas and depressions. The mounded areas, which might represent the collapse of individual structures, have the same characteristics as low mounded sites on CORONA imagery. The depressions represent extraction pits for mud brick material; they accumulate moisture, and in some seasons vegetation as well, and therefore appear as dark areas (Fig. 3.1c).

These reflectance signatures are contingent on ground conditions, particularly soil moisture and vegetation. In the Mediterranean climates characteristic of much of the Near East, rains fall in the winter, and crops reach their height in the late spring. Therefore, CORONA imagery from December through May have proven to be the most sensitive to anthropogenic soils. In summer and early fall imagery, crops have been harvested and the summer heat has removed most soil moisture. In these scenes, high mounds still cast shadows, but anthropogenic soils are obscured by the overall dryness and de-vegetated state of the landscape. In irrigated areas, such as southern Iraq, seasonal visibility is tied to the irrigation calendar.

The utility of CORONA for site identification has been recognized by archaeologists, especially in exploring site elements not easily seen on the ground. Studies have considered the lower towns of the cities of Samosata (Samsat) and Carchemish on the Turkish Tigris (Kennedy 1998) and the capitals of the Assyrian empire (Scardozzi 2011). The hinterland of the Early Islamic city of Raqqa on the Syrian Euphrates has been documented in great detail, including roads, tracks, canals, industrial areas, and even previously unrecognized palaces (Challis et al. 2002–2004; Challis 2007). CORONA allows virtual survey of regions where ground observation would be difficult or impossible. For example, CORONA-based analysis of southern Mesopotamia (Hritz 2004, 2010) has nearly doubled the number of sites identified by the surveys of Robert McC. Adams and colleagues; a similar reassessment has been undertaken for the core of the Assyrian empire in northern Mesopotamia (Altaweel 2008).

Despite its great potential, the systematic use of CORONA to guide field survey has been less common, but where attempted, it has proven to be highly effective. For example, in the Homs region of western Syria, CORONA resolved mounds, landscape features, and even sherd scatters (Philip et al. 2002; Wilkinson et al. 2006; Beck et al. 2007; Philip and Bradbury 2010). In the Amuq Valley of southeastern Turkey, all known sites could be identified in CORONA, and a host of previously unrecognized ones added to the map (Casana and Wilkinson 2005; Casana 2007). Finally, a survey of 125 km^2 around the Early Bronze Age city of Hamoukar in northeastern Syria identified 60 archaeological sites (Ur 2010). Pre-survey CORONA analysis had already identified all of them. A program of transect walking failed to recover any additional sites not previously recognized remotely. These sites range from tiny 0.2 ha mounds to a massive 300 ha low density complex (Ur 2010: 53–57; Al Quntar et al. 2011), the full extent of which would not have been recognized without CORONA photographs.

3.3 "Off-site" Landscape Features

Habitation sites are only one component of past landscapes, albeit a highly visible one. The most significant result of CORONA-based analyses has been the recovery of "off-site" landscape elements.

3.3.1 Roads and Tracks

In northern Mesopotamia, sites of the Early Bronze Age (ca. 2600–2000 B.C.), village and city alike, were connected in an extensive network of linear trackways (Ur 2003, 2009, 2010: 76–87, 129–146; Wilkinson et al. 2010). Their depressed morphologies collect moisture and appear as dark lines. As with settlement sites, their visibility is closely connected to land use and ground conditions; for this

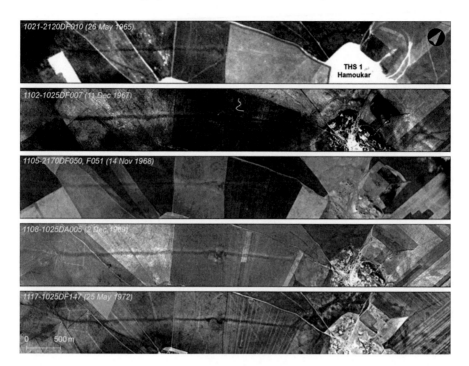

Fig. 3.2   Early Bronze Age trackway to Hamoukar, northeastern Syria

reason, it is useful to have a range of scenes from different times in order to fully
trace the surviving elements (Fig. 3.2). Ultimately more than 6,000 km of pre-modern
trackways were mapped, most datable via association to the Early Bronze Age, but
with a small Early Islamic (ca. A.D. 600–900) component. Morphologically similar
trackways have been identified on CORONA along the Turkish Tigris (Kennedy
1998) and in northern Iraq (Altaweel 2003, 2008).

3.3.2   Irrigation and Water Management

Pre-modern irrigation is particularly well suited to CORONA-based analysis. The
channel bed of a former canal is often dark with lighter margins that represent the
backdirt from their excavation and the upcast from maintenance. In the largest
canals, these spoil banks may cast shadows on their northern sides.

In northern Iraq, an extensive network of canals was constructed by the kings of
the Assyrian empire. These features had been observed anecdotally over the last
century and a half, but not yet investigated systematically on the ground.
Comprehensive remote sensing surveys (Ur 2005; Altaweel 2008: 72–77) have now
located several other features that match the signature of known canals, and added

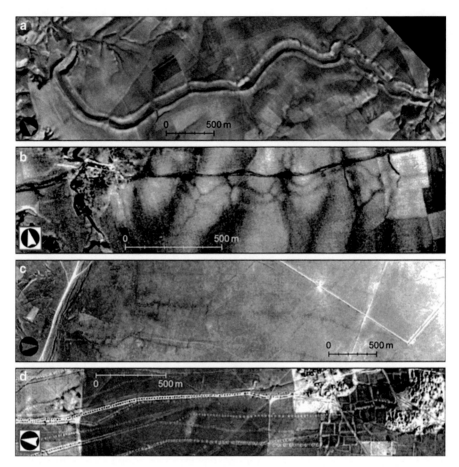

Fig. 3.3 Irrigation features. (**a**) Major Assyrian earthwork near Bandwai, Iraq (1108-1025DA007, 2 Dec 1969); (**b**) Assyrian canal with local offtakes near Jerwan, Iraq (1039-2088DA033, 28 Feb 1967); (**c**) Sasanian irrigation in the Mughan Steppe, NW Iran (1103-1057DF074, 5 May 1968); (**d**) *Qanat* shafts near Sinjar, northern Iraq (1102-1025DF012, 11 Dec 1967)

important details (Fig. 3.3a). For instance, most interpretations of the Assyrian system above Nineveh assume that all water went to the imperial capital. A reassessment using CORONA shows abundant evidence for off-takes throughout the system, such as in the region of the famous aqueduct at Jerwan (Fig. 3.3b). Without ground survey, however, the beneficiaries of these local irrigation offtakes remain unknown.

Another imperial irrigation system could be documented on the Mughan steppe of northwestern Iran (Alizadeh and Ur 2007: 151–154). At the time of image acquisition, the high banks of the primary canal cast shadows, and up to four levels of distributaries below it could be discerned (Fig. 3.3c). All are dated to the period of the Sasanian empire (ca. A.D. 224–651) on the basis of their association with fortified structures, a dating which makes the system contemporary with similar irrigation systems elsewhere in the Sasanian empire (Wilkinson and Rayne 2010: 122; Omrani et al. 2007). Today, all but the fortified sites have been obliterated by twentieth century A.D. irrigation.

Other irrigation systems have been studied purely via CORONA and GAMBIT, or anecdotally. A 300 km^2 irrigation system below Nisibin in Syria could be mapped in its entirety via GAMBIT (Ur 2010: 137–138). It was probably in use in the early Ottoman period (sixteenth century A.D.) but may have originated as early as the first millennium B.C.; it awaits systematic exploration. A particularly ripe direction for future research lies in the origins and development of subterranean *qanat* or *karez* systems. *Qanat* shafts appear as dark spots with lighter rings around them (Fig. 3.3d). They have been noted on CORONA around Islamic Raqqa (Challis et al. 2002–2004) and elsewhere in the Near East (Wilkinson and Rayne 2010). A preliminary inspection of CORONA for the plain around Erbil has recognized almost 7,000 shafts over 400 km^2 around the city; these features were a focus of field survey starting in 2012.

3.3.3 Field Systems

In alluvial areas of the Near East, fields are demarcated by low earthen ridges that are easy obliterated; few if any such pre-modern systems have survived to the present. In rocky areas, field clearance led to more durable stone walls that can fall within the limits of CORONA's resolution. In the basaltic "sub-optimal" area east of Homs, Syria, it was possible to recognize a palimpsest of stone-walled field systems that probably originated with Roman centuriation, along with cairn fields and stone enclosures (Philip et al. 2002; Philip and Bradbury 2010). Elsewhere, CORONA imagery has been used to study field systems only indirectly. For example, the terminal ends of hollow ways mark the interface between outfields and pasture (Ur and Wilkinson 2008), and irrigation canals can also offer clues (Alizadeh and Ur 2007: 153).

3.4 Nomadic Landscapes

For the most part, the landscape traces of mobile groups are too ephemeral for CORONA-based analysis, with at least one exception. Since the seventeenth century A.D., Shahsevan nomads wintered on the Mughan steppe of northwestern Iran (Fig. 3.4). On CORONA scenes, their campsites appear as light circles that can be mapped in detail (Alizadeh and Ur 2007: 154–157). A full coverage examination revealed significant patterning. Today, campsite remains are only found in the upland region of the steppe, and are absent on the river terraces. In the

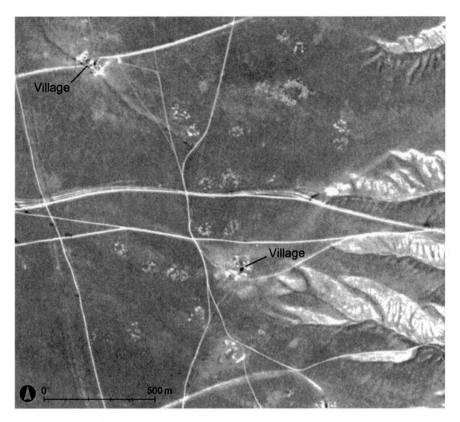

Fig. 3.4 Pastoral nomadic campsites of the Shahsevan tribal confederacy, Mughan Steppe, NW Iran (1110-1057DA111, 24 May 1970)

archaeological landscape of 40 years ago, however, campsite remains occurred throughout the steppe at comparable densities. The absence of campsite remains from the terraces is clearly a taphonomic pattern, and not related to the exclusion, voluntary or otherwise, of nomads from agricultural lands.

3.5 Landscapes and Environments

The most extensive work on past environments has been undertaken in regions of southern Iraq, where ground-based work has been impossible for two decades (Fig. 3.5). For example, Hritz (2010) has proposed a location for the Tigris River in the fifth-third millennia B.C., based on sites and meander scars on CORONA scenes, as well as SRTM terrain data. Her analysis resolves a longstanding conflict between archaeological and textual data on the river's position, and highlights the dynamism of the Tigris and its leveed branches.

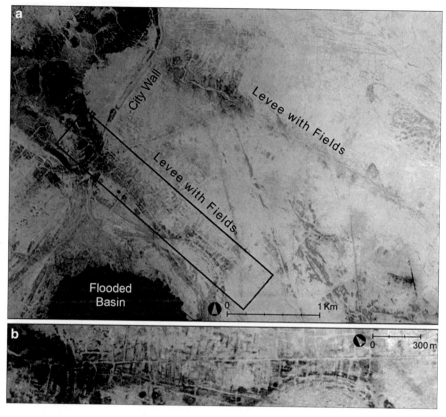

Fig. 3.5 Southeastern hinterland of the city of Uruk, southern Iraq (1103-1041DA059, 4 May 1968). (**a**) Relict river levees with irrigation channels and field systems atop them; (**b**) Detail showing meander scars, channels, and field systems

CORONA data underpins a recent challenge to the agricultural basis of early Sumerian urban civilization. Pournelle (2003, 2007) reassessed the locations of fifth and fourth millennium B.C. sites in southern Iraq on the basis of CORONA imagery and geomorphological data. Early sites often sat atop "turtlebacks," isolated relict terraces of Pleistocene date, and that they were associated with "bird's foot" deltas (Pournelle 2007: 43–45). Both of these features are characteristic of marsh environments. Pournelle concludes that the economic foundations of initial urbanism were marsh resources like fish and reeds, rather than the agro-pastoral systems that supported later cities.

3.6   Conclusions

Imagery from CORONA and related intelligence programs have truly revolution-ized the study of landscapes in the Near East. They have not, however, been exploited to their full potential. There has been a tendency by archaeologists to assess a single scene, and when features are not immediately apparent, to dismiss the utility of CORONA generally. It must be stressed that scenes from multiple missions, taken under different ground conditions, should be inspected (see, e.g., Fig. 3.2). The extensive geographic coverage of CORONA has also been under-exploited. This review has focused on broad-area systematic investigations of sites and features (canals, tracks, etc.), but many more studies have focused anecdotally on individual sites or landscape features. This situation may change soon with the launch of the online *CORONA Atlas of the Middle East* (Casana and Cothren, this volume).

There are substantial limitations to these CORONA-based approaches. They are most successful in alluvial environments where earthen architecture predominated. Furthermore, they work best with sedentary habitation sites. Finally, their use cannot eliminate the critical stage of ground observation and systematic artifact collection. Generally, however, there is little reason, beyond imagery availability, why a CORONA-based approach would not be exportable to sedentary alluvial landscapes across the planet.

Acknowledgements This manuscript was strengthened by critical comments from Li Min, Carrie Hritz, Fred Limp, and Michael Harrower.

References

Al Quntar, S., Khalidi, L., & Ur, J. A. (2011). Proto-urbanism in the late 5th millennium BC: Survey and excavations at Khirbat al-Fakhar/Hamoukar, Northeast Syria. *Paléorient, 37*(2), 151–175.

Alizadeh, K., & Ur, J. A. (2007). Formation and destruction of pastoral and irrigation landscapes on the Mughan Steppe, North-Western Iran. *Antiquity, 81*(311), 148–160.

Altaweel, M. (2003). The roads of Ashur and Nineveh. *Akkadica, 124*(2), 221–228.

Altaweel, M. (2008). *The imperial landscape of Ashur: Settlement and land use in the Assyrian heartland (Heidelberger Studien zum Alten Orient 11)*. Heidelberg: Heidelberger Orientverlag.

Beck, A., Philip, G., Abdulkarim, M., & Donoghue, D. (2007). Evaluation of Corona and Ikonos high resolution satellite imagery for archaeological prospection in Western Syria. *Antiquity, 81*, 161–175.

Casana, J. (2007). Structural transformations in settlement systems of the Northern Levant. *American Journal of Archaeology, 111*(2), 195–221.

Casana, J., & Wilkinson, T. J. (2005). Settlement and landscapes in the Amuq Region. In K. A. Yener (Ed.), *The Amuq Valley regional projects, volume 1: Surveys in the plain of Antioch and Orontes Delta, Turkey, 1995–2002* (pp. 25–66, Oriental Institute Publications 131). Chicago: Oriental Institute.

Challis, K. (2007). Archaeology's cold war windfall: The CORONA programme and lost landscapes of the Near East. *Journal of the British Interplanetary Society, 60*(1), 21–27.

Challis, K., Priestnall, G., Gardner, A., Henderson, J., & O'Hara, S. (2002–2004). Corona remotely-sensed imagery in dryland archaeology: The Islamic city of al-Raqqa, Syria. *Journal of Field Archaeology, 29*, 139–153.

Fowler, M. J. F. (2006). Modelling the acquisition times of CORONA KH-4B satellite photographs. *AARGnews, 33*, 34–39.

Hritz, C. (2004). The hidden landscape of southern Mesopotamia. *Akkadica, 125*, 93–106.

Hritz, C. (2010). Tracing settlement patterns and channel systems in southern Mesopotamia using remote sensing. *Journal of Field Archaeology, 35*(2), 184–203.

Kennedy, D. (1998). Declassified satellite photographs and archaeology in the Middle East: Case studies from Turkey. *Antiquity, 72,* 553–561.

Kouchoukos, N. (2001). Satellite images and Near Eastern landscapes. *Near Eastern Archaeology, 64,* 80–91.

McDonald, R. A. (Ed.). (1997). *Corona between the Sun and the Earth: The first NRO reconnaissance eye in space.* Bethesda: American Society for Photogrammetry and Remote Sensing.

Omrani, H., Sauer, E. W., Wilkinson, T. J., Tamak, E. S., Ainslie, R., Mahmoudi, M., et al. (2007). An imperial frontier of the Sasanian empire: Further fieldwork at the Great Wall of Gorgan. *Iran, 45,* 95–136.

Philip, G., & Bradbury, J. (2010). Pre-classical activity in the basalt landscape of the Homs region, Syria: Implications for the development of 'sub-optimal' zones in the Levant during the Chalcolithic-Early Bronze Age. *Levant, 42*(2), 136–169.

Philip, G., Dononghue, D., Beck, A., & Galiatsatos, N. (2002). CORONA satellite photography: An archaeological application from the Middle East. *Antiquity, 76,* 109–118.

Poidebard, A. (1934). *La trace de Rome dans le désert de Syrie (Bibliothéque Archéologique et Historique 18).* Paris: Librairie Orientaliste Paul Geuthner.

Pournelle, J. R. (2003). The littoral foundations of the Uruk state: Using satellite photography toward a new understanding of 5th/4th millennium BCE landscapes in the Warka Survey Area, Iraq. In D. Gheorghiu (Ed.), *Chalcolithic and Early Bronze Age hydrostrategies* (pp. 5–23, BAR International Series 1123). Oxford: Archaeopress.

Pournelle, J. R. (2007). KLM to Corona: A bird's eye view of cultural ecology and early Mesopotamian urbanization. In E. C. Stone (Ed.), *Settlement and society: Essays dedicated to Robert McCormick Adams* (pp. 29–61). Los Angeles: Cotsen Institute of Archaeology.

Richelson, J. (2003). A 'Rifle' in space. *Air Force Magazine, 86,* 72–75.

Scardozzi, G. (2011). Multitemporal satellite images for knowledge of the Assyrian capital cities and for monitoring landscape transformations in the upper course of the Tigris River. *International Journal of Geophysics,* Article ID 917306 (2011). doi:10.1155/2011/917306.

Ur, J. A. (2003). CORONA satellite photography and ancient road networks: A Northern Mesopotamian case study. *Antiquity, 77,* 102–115.

Ur, J. A. (2005). Sennacherib's northern Assyrian canals: New insights from satellite imagery and aerial photography. *Iraq, 67*(1), 317–345.

Ur, J. A. (2009). Emergent landscapes of movement in Early Bronze Age Northern Mesopotamia. In J. E. Snead, C. Erickson, & W. A. Darling (Eds.), *Landscapes of movement: Paths, trails, and roads in anthropological perspective* (pp. 180–203). Philadelphia: University of Pennsylvania Museum Press.

Ur, J. A. (2010). *Urbanism and cultural landscapes in northeastern Syria: The Tell Hamoukar survey, 1999–2001* (Oriental Institute Publications 137). Chicago: University of Chicago Oriental Institute.

Ur, J. A., & Wilkinson, T. J. (2008). Settlement and economic landscapes of Tell Beydar and its hinterland. In M. Lebeau & A. Suleiman (Eds.), *Beydar studies I* (pp. 305–327). Turnhout: Brepols.

Wilkinson, K. N., Beck, A. R., & Philip, G. (2006). Satellite imagery as a resource in the prospection for archaeological sites in central Syria. *Geoarchaeology, 21*(7), 735–750.

Wilkinson, T. J., & Rayne, L. (2010). Hydraulic landscapes and imperial power in the Near East. *Water History, 2,* 115–144.

Wilkinson, T. J., French, C., Ur, J. A., & Semple, M. (2010). The geoarchaeology of route systems in northern Syria. *Geoarchaeology, 25*(6), 745–771.

Chapter 4
The CORONA Atlas Project: Orthorectification of CORONA Satellite Imagery and Regional-Scale Archaeological Exploration in the Near East

Jesse Casana and Jackson Cothren

Abstract Declassified, Cold War-era CORONA satellite images have proven to be a critical tool in archaeology of the Near East and elsewhere, primarily because they preserve a picture of sites and landscapes that predates recent agricultural, industrial and urban development. However, unprocessed CORONA images contain extreme spatial distortions caused by a cross-path panoramic scanning system, and the absence of detailed orientation and camera information makes correction of these errors highly challenging, resulting in small-scale, piecemeal application of this resource. This chapter overviews our methods for efficient orthorectification of KH-4A and KH-4B CORONA imagery and our development of a freely accessible online database for viewing and distribution of corrected images. We also highlight our efforts to deploy regional-scale CORONA coverage to facilitate the discovery and documentation of archaeological landscapes from across the Northern Fertile Crescent.

Keywords CORONA • Orthorectification • Near East • Settlement history • Site morphology

4.1 Introduction

Archaeologists have long appreciated the extraordinary power of aerial photography and satellite imagery to aid in the discovery and interpretation of archaeological sites and ancient cultural features. However, in the Middle East, little imagery of adequate spatial resolution was available to archaeologists until 1995, when a large archive of US intelligence satellite images collected by the CORONA program from 1960 to 1972 was declassified and made publicly available (Day et al. 1998). Because CORONA images are more than 30 years old, they preserve a picture of archaeological landscapes prior to their destruction by recent industrialization, urban expansion and agricultural intensification, making this imagery an absolutely unique resource that cannot be replaced by new technologies. CORONA satellite

D.C. Comer and M.J. Harrower, *Mapping Archaeological Landscapes from Space*, SpringerBriefs in Archaeology, DOI 10.1007/978-1-4614-6074-9_4, © Springer Science+Business Media New York 2013

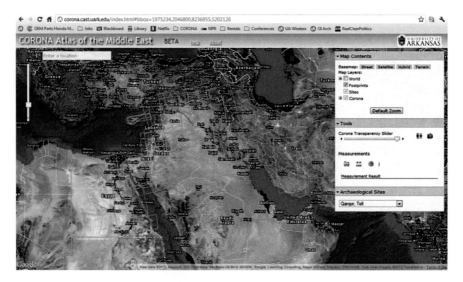

Fig. 4.1 CORONA Atlas, showing extent of orthorectified images as of April 2012

imagery, especially imagery from the KH-4A to KH-4B series of missions, has now been established as an invaluable resource in archaeology of the Near East and elsewhere (e.g., Beck et al. 2007; Casana 2007; Casana and Wilkinson 2005; Casana and Cothren 2008; Casana et al. 2012; Challis et al. 2004; Fowler 2004; Kennedy 1998; Kouchoukos 2001; Philip et al. 2002; Ur 2003, 2005, 2010, this volume).

However, the raw, unprocessed images that are provided by the United States Geological Survey (USGS) contain extreme spatial distortions caused by the cross-path panoramic scanning system, such that integrating the images into GIS databases is extremely challenging because no commercial software packages offer solutions for the unusual CORONA image format. We have now developed more efficient means of orthorectifying CORONA imagery and have used these techniques to correct more than 1,200 images of the Middle East and surrounding areas (Fig. 4.1), all of which can now be freely viewed and downloaded online (http://corona.cast.uark.edu/index). This paper briefly outlines our methods for geometric correction of CORONA imagery and presents preliminary results of an ongoing effort to explore the archaeological landscape visible in our imagery database (also see, Casana et al. 2012).

4.2 Geometric Correction of CORONA Imagery

The highest resolution CORONA images, produced by the KH-4B generation of satellites in operation from 1967 to 1972, were collected by a panoramic camera on long film strips, each covering 8.6 × 117 km of the ground with a nadir resolution of approximately 1.83 m. The system included both a forward and aft camera, so that highly-convergent stereo imagery (30°) was collected for most areas. The panoramic camera system, while effective at covering large areas at high spatial resolution, resulted in severe geometric distortion on the images. Moreover, unlike modern satellite imaging systems, no detailed positional or orientation information is available, and many camera parameters remain unknown, making automated geometric correction impossible. Until now, most researchers who attempted to correct the spatial distortions contained in CORONA imagery have relied on two-dimensional curve fitting techniques such as higher order polynomials or a variety of spline-based methods (the techniques are often colloquially referred to as rubber-sheeting because of their effect of differentially stretching and warping the image to fit the ground control). Because the two-dimensional methods cannot take into account distortions caused by the three-dimensional imaging geometry and terrain, the process is labor intensive, requires many ground control points, and is only reasonably accurate within small areas. In fact, it often results in large spatial errors, especially in high-relief areas, and always prohibits stereo analysis or DEM extraction.

For researchers focusing on relatively small areas (c. 100 km²), orthorectification can be achieved by extracting a small image segment from a larger CORONA scene (<15%) and treating it as a simple frame camera, as described in our earlier work (Casana and Cothren 2008) and similar to that undertaken by other researchers (Altmaier and Kany 2002; Galiatsatos et al. 2011). While reasonably effective over small areas, this procedure is time-consuming owing to the large number of ground control points (GCPs) that must be collected from another ortho-image and produces increasingly poor results towards the edges of imagery where panoramic distortions are greatest.

Our more recent work, described in Casana et al. (2012) has instead developed a method for efficient orthorectification of full CORONA scenes. Because the USGS provides CORONA as four overlapping scanned segments, we first reconstruct complete CORONA scenes using an automated scale-invariant feature transform (SIFT) algorithm (Lowe 2004). Pseudo-fiducials are established at the corners of individual images, and overlapping scenes derived from a single satellite revolution are assembled into large blocks. We then collect 10–20 GCPs per scene, focusing on areas of overlap between adjacent images, by manually measuring the image coordinates of features visible on CORONA in ERDAS' Leica Photogrammetry Suite, deriving horizontal coordinates from Google Maps reference data using a custom-built digitizer tool (Fig. 4.2a). Elevations at these locations are then interpolated from SRTM Level 1 data.

Next, we perform a block bundle adjustment to minimize the reprojection error between the observed and predicted location of image points, finding the best fit between images and GCPs (Fig. 4.2b). Following a solution proposed by Sohn et al. (2004) we employ modified collinearity equations to model panoramic distortions. A physical sensor model is then created for each CORONA scene in which camera parameters and ephemeris information are estimated by the panoramic sensor model in a least squares process. After these parameters are estimated, they can be used to

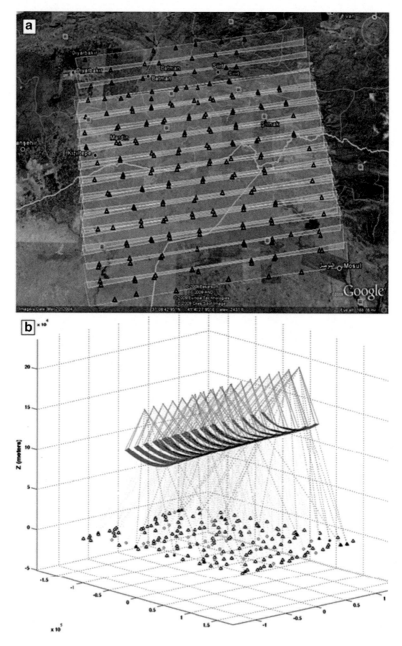

Fig. 4.2 (**a**) Collection of ground control points from Google Maps image data on a single block of CORONA images in eastern Turkey; (**b**) Orientation of CORONA images for the same block

map ground points (latitude, longitude, and height) to a particular pixel in the image. Because no commercial GIS is able to make direct use of these parameters and display the image correctly, we then map these unusable parameters to a well-known set of general parameters known as Rational Polynomial Coefficients (RPCs). Since all commercial satellite vendors use the same set of RPCs, virtually all GIS software with basic image processing capabilities can project the RPC-enabled CORONA images to a ground plane or DEM.

Images are stored and distributed as National Imagery Transfer Format (NITF) files, preserving the original image, parameters used to geometrically correct the image, as well as other image metadata. Orthorectification of NITF images in our online database is achieved by projecting images over an SRTM digital elevation model. Accuracy of most of our orthorectified images range from 3–10 m at nadir to 20–80 m at the edges. Users requiring greater accuracy can input a small number of additional ground control points or utilize a better digital elevation model than the SRTM upon which we rely. The rigorous processes we employ also enable CORONA NITFs to be easily viewed in stereo or used for the extraction of digital elevation data using a variety of softcopy photogrammetry software packages (Casana and Cothren 2008; Casana et al. 2012).

4.3   Exploring Regional-Scale CORONA in the Northern Fertile Crescent

The clarity with which sites and features can be resolved on CORONA imagery across much of the Near East (Ur, this volume) has revolutionized archaeological survey projects where these data have been employed, as in the Amuq Plain of southern Turkey (Casana and Wilkinson 2005; Casana 2007) the Jazireh region of eastern Syria (Ur 2010; Wright et al. 2006–2007), and the Homs area of western Syria (Beck et al. 2007; Philip et al. 2002). However, to date such studies have been undertaken in a piecemeal way, focused on relatively small study areas that were the subject of survey or excavation projects. Moreover, the vast majority of past surveys did not have access to any aerial imagery, and today much of the Near East been heavily impacted by agricultural and urban development. Using the orthorectification methods described above, we have produced systematic CORONA coverage for much of the Near East offering an unprecedented opportunity to virtually survey vast areas, beyond project boundaries and across national borders. With support from the NASA Space Archaeology Program, our recent work focuses on a study area of approximately 300,000 km^2 in the northern Fertile Crescent, encompassing northern Lebanon, Syria, southern Turkey and northern Iraq, and seeks to map all sites and features visible on CORONA across the region.

4.3.1   Revisiting Previous Surveys

The first objective in our project has been to assemble as many previously published survey projects as possible and to map sites those projects recorded. Analysis of the 35-plus surveys often reveals the richness and complexity of archaeological landscapes that went unrecorded by earlier researchers. For example, in the Qoueiq River Valley north of Aleppo, Syria, a survey project in the late 1970s attempted to record as many sites as possible (Matthers 1981), but low-intensity methods inevitably missed many sites and features in the region. Dozens of sites can be seen on CORONA imagery between those ones discovered and mapped by the project, highlighting how little we may know about settlement history in many parts of the Near East.

Even at sites that were recorded, much can be learned through analysis of CORONA. For example, the Qoueiq survey includes the major mounded site of Tell Rifa'at, almost certainly the ancient city of Arpad, well-known as one of the pre-eminent powers of the region during the Iron Age in the early first millennium B.C. While Rifa'at was recorded by the Qouieq survey and briefly excavated, the existence of a massive lower city surrounding the high citadel at the center of the site went undocumented (Fig. 4.3a). This lower town, demarcated by a near-perfectly circular fortification wall enclosing some 120 ha, has now been largely obscured by the modern town, but its recognition on CORONA reveals that the ancient city of Arpad was perhaps the largest pre-Classical city in the Levant. Lower towns like that at Tell Rifa'at have now been recorded at dozens of sites across the study area (Casana et al. 2012).

4.3.2   Discovering New Sites

Beyond the bounds of previous survey projects, there are vast areas of the Near East that have never been systematically investigated by archaeologists, and in these unrecorded regions there are remarkable sites that are still undocumented. For example, in the Araban Plain just west of the Euphrates River in south-central Turkey, CORONA imagery reveals the sprawling, 50 + ha mound of Araban Höyük, with a high, central citadel and a large lower town, today mostly covered by the modern city (Fig. 4.3b). Despite the size and prominence of the site, it remains virtually unknown to archaeologists. Tens of thousands of other small sites, such as the small mound shown in Fig. 4.3c, similarly remain unrecorded and, using our database of known sites as a guide, our project is working systematically to map all site-like features across our study area.

4.3.3   Documenting Landscape Features

As with sites themselves, ancient landscape features such as roads, canals and field systems are often visible on CORONA imagery, preserved as soil marks, crop

Fig. 4.3 Comparisons between 1968 CORONA and modern commercial satellite images for (**a**) Tell Rifa'at, western Syria, where a large, circular city wall is visible (**b**) Araban Höyük, southern Turkey, a still undocumented 60 ha fortified site, and (**c**) a small, unrecorded tell in the Mardin Plain of eastern Turkey

marks, or subtle topographic features in the landscape (Ur, this volume). These ephemeral features are among the most vulnerable to destruction by modern land use. For example, the distinctive systems of radial roadways surrounding mounded sites across the northern Mesopotamian plains have been well documented using CORONA imagery (Ur 2003), although most of these features have been erased or obscured by modern land use. Most previous research has concluded that these ancient roads, preserved only as subtle depressions in the landscape, are almost exclusively associated with settlements of the third millennium B.C., as at Tell Brak (Fig. 4.4a), and located entirely within a narrow zone through eastern Syria and

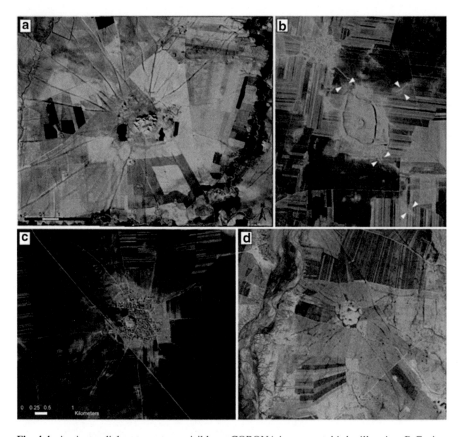

Fig. 4.4 Ancient radial route systems visible on CORONA imagery at third millennium B.C. sites of (**a**) Tell Brak, eastern Syria, and (**b**) Tell Mardikh (Ebla), western Syria. We have found much later examples dating to the early first millennium B.C. at (**c**) Tell Rifa'at, northwestern Syria, and to the early medieval period at (**d**) KS107, southwestern Iran

northern Iraq. Our recent analysis of regional-scale CORONA imagery has now documented very similar features across a much larger region, from the eastern Mediterranean to southwestern Iran and associated with sites of much later periods, including at the first millennium B.C. site of Tell Rifa'at (Fig. 4.4c), at Roman/late Roman sites in the Orontes Valley of western Syria, and at a major Sasasian/early Islamic site, KS107, in southwestern Iran (Fig. 4.4d; Casana forthcoming).

4.3.4 Mapping Regional Settlement Systems

The comprehensive database of archaeological sites and features we are building using regional-scale CORONA imagery offers many opportunities for analyses of settlement systems. One approach involves analysis of site morphology, which can be a coarse but reliable dating tool (Casana 2012). Many features visible on imagery,

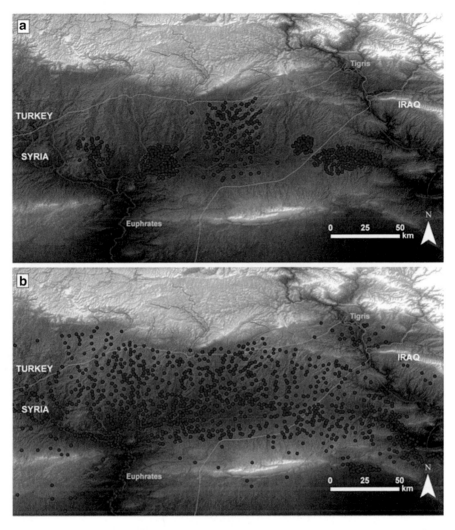

Fig. 4.5 (a) All sites recorded by published regional surveys in eastern Syria and northern Iraq; (b) All conical mounded sites visible on CORONA imagery in the same area, likely representing the full extent of the third millennium B.C. settlement system (image credit: Tuna Kalayci)

such as the presence and shape of mounding, rectilinear architecture, or lower towns that surround sites are frequently associated with settlements of a particular period within a given region. By systematizing these often causal observations, essentially treating sites as cultural artifacts, we can approximately date many sites based solely on their appearance. We can thus map settlement systems – or at least settlements of a particular type – across large regions.

For example, in the Jazireh region of eastern Syria and northern Iraq, regional archaeological surveys have demonstrated that virtually all prominent, conical mounded sites (e.g., Fig. 4.3c) were occupied during the mid to late third millennium B.C., while little evidence of occupation during that period can be found elsewhere (Ur 2010; Wright et al. 2006–2007). However, surveys have only been undertaken over a small percentage of the area, resulting in an incomplete picture of regional settlement (Fig. 4.5a). Fortunately, these third millennium B.C. mounds are relatively easy to recognize on CORONA imagery, meaning that they can readily be mapped beyond the limits of survey boundaries and across national borders. The resulting analysis (Kalayci forthcoming), mapping all such sites across the entire region, provides a better picture of the complete third millennium B.C. settlement system than would otherwise be possible (Fig. 4.5b).

Our ongoing research uses a similar approach to map other site-types characteristic of different periods and regions. The resulting geographic and temporal distribution of site-types are then evaluated against a range of environmental variables, including high-temporal resolution AVHRR-derived NDVI values and inter-annual precipitation patterns. More broadly, our work demonstrates the potential of regional-scale, high-resolution imagery such as CORONA to serve as a powerful tool for exploration of archaeological landscapes.

References

Altmaier, A., & Kany, C. (2002). Digital surface model generation from CORONA satellite images. *ISPRS Journal of Photogrammetry and Remote Sensing, 56*(4), 221–235.

Beck, A., Philip, G., Abdulkarim, M., & Donoghue, D. (2007). Evaluation 17 of Corona and Ikonos high resolution satellite imagery for archaeological prospection in western Syria. *Antiquity, 81*(1), 161–175.

Casana, J. (2007). Structural transformations in settlement systems of the northern Levant. *American Journal of Archaeology, 111*, 195–221.

Casana, J. (2012). Site morphology and settlement history in the northern Levant. In R. Matthews, J. Curtis, M. Seymour, A. Fletcher, A. Gascoigne, C. Glatz, S. J. Simpson, H. Taylor, J. Tubb, R. Chapman (Eds.), *Proceedings of the 7th International Congress of the Archaeology of the Ancient Near East (7th ICAANE)*, 12–16 Apr 2010 (pp. 593–608). London: The British Museum & University College London Press.

Casana, J. (forthcoming). Radial route systems and agro-pastoral strategies in the ancient Near East: New discoveries from western Syria and southwestern Iran. *Journal of Anthropological Archaeology.*

Casana, J., & Cothren, J. (2008). Stereo analysis, DEM extraction and orthorectification of CORONA satellite imagery. *Antiquity, 82*, 732–749.

Casana, J, & Wilkinson, T. J. (2005). Settlement and landscapes in the Amuq region. In K. A. Yener (Ed.), *The Amuq Valley Regional Projects, Volume 1, Surveys in the Plain of Antioch and Orontes Delta, Turkey, 1995–2002* (pp. 25–65, Oriental Institute Publication No. 131). Chicago: University of Chicago Press.

Casana, J., Cothren, J., & Kalayci, T. (2012). Swords into ploughshares: Archaeological applications of CORONA satellite imagery in the Near East. *Internet Archaeology, 32*(2).

Challis, K., Priestnall, G., Gardner, A., Henderson, J., & O'Hara, S. (2004). CORONA remotely-sensed imagery in dryland archaeology: The Islamic city of al-Raqqa, Syria. *Journal of Field Archaeology, 29*, 139–153.

Cothren, J., Casana, J., Barnes, A., & Kalayci, T. (forthcoming). *Agricultural Production Stabiltity of Settlement Systems in Northern Mesopotamia during the Early Bronze Age (third millennium BCE)*. An efficient method for rigorous orthorectification of CORONA satellite imagery. ISPRS Journal of Photogrammetry and Remote Sensing.

Day, D. A., Logsdon, J. M., & Latell, B. (1998). *Eye in the sky: The story of the CORONA spy satellites*. Washington, DC: Smithsonian Institution Press.

Fowler, M. J. (2004). Declassified CORONA-KH4B satellite photography of remains from Rome's desert frontier. *International Journal of Remote Sensing, 25*(18), 3549–3554.

Galiatsatos, N., Donoghue, D., & Philip, G. (2011). High resolution elevation data derived from stereoscopic CORONA imagery with minimal ground control: An approach using Ikonos and SRTM data. *Photogrammetric Engineering and Remote Sensing, 74*(9), 1093–1106.

Kalayci, T. (forthcoming). Ph. D. dissertation, Department of Anthropology, University of Arkansas. Agricultural Production Stability of Settlement Systems in Northern Mesopotamia during the Early Bronze Age (thrid millennium BCE).

Kennedy, D. (1998). Declassified satellite photographs and archaeology in the Middle East: Case studies from Turkey. *Antiquity, 72,* 553–561.

Kouchoukos, N. (2001). Satellite images and the representation of Near Eastern landscapes. *Near Eastern Archaeology, 64*(1–2), 80–91.

Lowe, D. G. (2004). Distinctive image features from scale-invariant keypoints. *International Journal of Computer Vision, 60,* 91–110.

Matthers, J. (Ed.). (1981). *The River Qoueiq, Northern Syria, and its catchment: Studies arising from the Tell Rifa'at survey 1977–79* (BAR. International Series 98). Oxford: BAR.

Philip, G., Donoghue, D., Beck, A., & Galiatsatos, N. (2002). Corona satellite photography: An archaeological application from the Middle East. *Antiquity, 76,* 109–118.

Sohn, H. G., Kim, G., & Yom, J. H. (2004). Mathematical modeling of historical reconnaissance CORONA KH-4B imagery. *The Photogrammetric Record, 19*(105), 51–66.

Ur, J. A. (2003). CORONA satellite photography and ancient road networks: A northern Mesopotamian case study. *Antiquity, 77,* 102–115.

Ur, J. A. (2005). Sennacherib's northern Assyrian canals: New insights from satellite imagery and aerial photography. *Iraq, 67,* 317–345.

Ur, J. A. (2010). *Urbanism and cultural landscapes in northeastern Syria: The Tell Hamoukar Survey, 1999–2001* (Oriental Institute Publications, 137). Chicago: Oriental Institute of the University of Chicago.

Wright, H. T., Rupley, E. S. A., Ur, J. A., Oates, J., & Ganem, E. (2006–2007). Preliminary report on the 2002 and 2003 seasons of the Tell Brak sustaining area survey. *Les Annales Archéologiques Arabes Syriennes, 49–50,* 7–21.

Chapter 5
Archaeological Landscapes of China and the Application of Corona Images

Li Min

Abstract With its comprehensive coverage and open access, declassified Corona images provide high-resolution images of the Chinese landscape before its recent industrial boom. Its enormous potential for understanding landscape transformations is not fully realized. This paper explores ways to incorporate Corona images into research and education on the archaeology of China. Examples drawn from three locations in the Qufu region are provided to illustrate the dynamic relationships between landscape and society.

Keywords China • Corona • Qufu • Landscape archaeology • Remote sensing

5.1 Introduction

As a rich legacy of the Cold War era, declassified Corona images have the distinctive advantage in representing archaeological patterns that are difficult to recognize at the ground level, e.g. ancient river channels, large rammed earth architectural foundations, and cultural modifications of the landscape. Corona images provide reasonably high-resolution images with a large coverage area (on single sheet and overall) from four decades ago at a low cost. These benefits make the Corona images a critical asset for teaching and research of Chinese archaeology. This paper explores preliminary efforts for incorporating Corona images into the field.

5.2 Teaching Chinese Archaeology with Corona Images

Teaching and research enhanced with remote sensing images has enormous prospects for understanding the social reconfiguration of historical landscapes in early civilizations. At the same time, the proliferation of new spatial technologies and the

D.C. Comer and M.J. Harrower, *Mapping Archaeological Landscapes from Space*, SpringerBriefs in Archaeology, DOI 10.1007/978-1-4614-6074-9_5, © Springer Science+Business Media New York 2013

archaeological data also take into account practical limitations such as the competing demands on time and resources. If a student takes an introductory course that offers both a survey of regional archaeology and an opportunity to conduct original research on spatial patterns with remote sensing data, s/he needs to strike a balance between the time invested in learning about the digital technology (e.g. GIS) and the time spent on mastering the cultural content.

The solution is a combination of Cold War legacy imagery with the open technology of our time, which offers a low cost, high resolution, and easily accessible approach to the spatial patterns of the past societies. For students who do not have GIS proficiency for advanced data analysis, their minimalist 'tool kit' would comprise of high resolution images from contemporary, open access sources such as Google Earth, Google Maps, Tianditu, and recently declassified images from the Corona program of intelligence satellite (Ur 2003, 2006). The effective use of this simple tool kit, however, could greatly enhance students' understanding of the past societies and may even at times allow them to make original contributions to existing research.

In order to take advantage of the spatial information provided by the declassification of Corona images and open source spatial data for archaeological education, I designed and offered an undergraduate course entitled 'Archaeological Landscapes of China.' Consisting of lectures and student independent projects involving Corona images, the experimental class was part of a university wide initiative for promoting the innovative use of emergent geo-technologies to transform the humanities pedagogy. This project involved the collaboration of Academic Technology Services, the W. M. Keck Center for Digital Humanities, and the Experiential Technologies Center at UCLA.

The lectures reviewed major benchmarks in the historical development of Chinese civilizations with an emphasis on the changes in spatial patterns associated with these social transformations, e.g. the rise of earliest cities and imperial capitals, construction of defense walls and monumental structures, building of the canals and irrigation systems, and configuration of sacred landscapes and pilgrimage sites. It allows students to place published archaeological works in new spatial contexts at various scales.

Each student worked on a class project by focusing on a region with major archaeological sites and conducted research on declassified high-resolution satellite images to identify spatial patterns of landscape transformation. Analysis of archaeological landscapes seen from the Corona images with high-resolution satellite images of the contemporary region provided by the Google Earth program presented students with diverse perspectives on the landscape features and a basis for developing archaeological interpretations. This combination provided an opportunity to conduct an original study on archaeological landscapes that have either vanished over the last three decades, or have become no longer accessible (Ur 2003; Wilkinson 2003). Through a combination of publication-based research and spatial analysis, the students developed new ways of visualizing and analyzing archaeological data.

5.3   Research Application of Corona Images in Chinese Archaeology

In addition to teaching, the research potential is equally promising. Corona imagery can make an important contribution to Chinese archaeology for several reasons. The timing of the Corona images is critical for site preservation. Taken at the height of the Cold War, these images captured the country amidst major campaigns of industrialized agricultural activities organized by the communes. In the process, mechanized deep plowing, land leveling, and the building of irrigation canal systems erased many surface features and monuments. For many areas, Corona images may be the only source of information on the landscape configurations before they were lost during this era of political tension.

A comparison of aerial photos taken over the Linzi region in 1938 and 1975 reveals that approximately 84 percent of ancient tomb mounds were lost over a period of four decades (Shandong Provincial Institute of Archaeology 2000: 34). While early aerial photos captured a landscape before industrialized land leveling, the coverage was rather limited and the resolution low. Similarly, while contemporary satellite images provide high-resolution images, destruction of the cultural landscape intensified from the industrial development and the urban expansion which taken together, compromised the quality of coverage. Much of the land surface is covered by industrial parks and plastic film for intensive agriculture, which offsets the benefit of increased resolution.

In practice, the research application of Corona images is dependent on the ecological zone and landscape conditions. The archaeological features in the pastoral highlands shared many characteristics with the Eurasian steppes and Near East, for which the analytical techniques discussed in other chapters of this book could be adopted. For the agrarian heartland, however, the landscape transformation from repeated occupation, agricultural intensification, flood sediments, and high population density make it very difficult to recognize the early archaeological features in China on remote sensing images.

An important step for understanding these transformations is to make the archaeological features "legible" by developing correlations between patterns observed from image analysis and archaeological field investigation. It involves a comparison with other lines of data, e.g. aerial photos, maps, geological reports, survey reports, excavation reports and, above all, actual field research. I will use three locales from the archaeological landscape of the Qufu region to illustrate these points (Fig. 5.1).

5.3.1   Bronze Age and Early Imperial City

The Bronze Age city of Qufu is one of the most important urban sites of early China. In 1977-78, archaeologist Zhang Xuehai and his team from the Shandong Provincial Institute of Archaeology systematically probed the urban site and mapped

Fig. 5.1 Map of the Qufu Region in Eastern China

out the subsurface distribution of palatial structures, administrative remains, craft production sites, and residential areas with an impressive 100,000 coring samples taken every five meters inside the rammed earth city wall. Subsequent excavations of Bronze Age cemeteries further provided evidence for cultural and social variations in material culture (Shandong Provincial Institute of Archaeology 1982).

With the help of this groundbreaking research, it is possible to compare the 1966 Corona image of Qufu with results from the probing survey and excavation conducted a decade later to identify urban features associated of the Bronze Age city (Fig. 5.2). Some of the prominent features include: the rammed earth wall, the moat surrounding the Bronze Age city, the early imperial palatial foundation in the center of the city, several historical temple complexes (represented by the black strips from concentration of cypress trees in courtyards), and a large historical cemetery north of the city site (National Bureau of Cultural Relics 2007: 383).

Chinese archaeology makes extensive use of subsurface probing to investigate underground foundations, which determine the presence or absence of rammed earth by allowing researchers to observe the variation in soil texture. The practice, however, is generally not extended to natural features, e.g. ancient river channels and lakes, which could be important for understanding the human habitat. These could be readily detectable from the Corona images due to differences in spectral reflectance.

By identifying these buried landscape features, new information can be gleaned from an analysis of Corona images even where some of these sites or features were already known and recorded from field surveys. For instance, dark, meandering striations on the land surface east of the city were good indicators of ancient river channels that had silted up during the historical period. How did the Bronze Age city manage its water system and physical terrain? How was the landscape transformed over time? These pose new questions for geomorphological and archaeological investigations. With the development of portable devices that integrate different

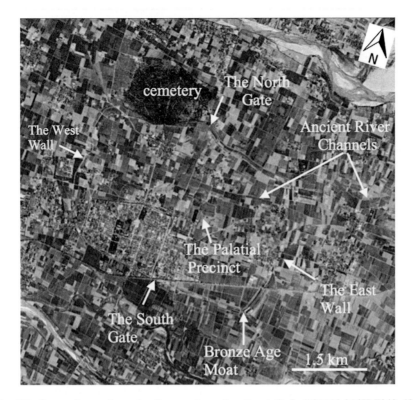

Fig. 5.2 Corona image from the Bronze Age city site at Qufu DS1115-2167DF040_40_c September 21, 1971 Longitude 116.99 Latitude 35.59

types of remotes sensing images and active GPS tracking, these questions can be addressed with real time use of Corona images in the field.[1]

Not all Chinese archaeological sites are as well preserved like the Bronze Age city at Qufu (Li 2006). Regional variations in landscape condition affect the contribution of Corona image for archaeology. In the lower Yellow River basin and the central Huai River basin, for instance, many historical sites were buried under thick layers of flood sediments from the Yellow River, which could measure up to ten meters in thickness, posing major challenges to remote sensing and surface survey (Jing et al. 1995; Rosen 2008). This is very different from the condition of investigating ancient tells and their surrounding road systems in Near East, where the impact of agriculture and urban formation is less severe.

However, rather than taking these transformations as obstacles for incorporating Corona images in the archaeological investigation in China, effective use of these images can help identify landscape changes that could potentially affect the

[1] I attribute this observation to Rachel Lee, a member of the Qufu project team.

visibility of archaeological features. Sediments filling in lakes and river channels often retain more moisture than the surrounding areas, thus are represented by darker color on Corona images. In the case of a regional survey, detecting these geomorphology issues from images and creating follow up investigations in the field could help prioritize investment in time and efforts.

5.3.2   The Royal Cemetery of the Lu State

Because of China's complex historical landscape, major changes in resolution could change within a short distance. In to the Qufu region, for instance, major landscape transformations around the area of the royal cemetery call for the effective use of Corona images from a very different perspective. The cemetery for the dukes of the Lu state from the middle first millennium B.C. was located on a ridge approximately 50 km west of the city. Textual records from the first millennium B.C. had identified the cemetery as the estate of the Lu nobles.

While the ancestral landscape of the Lu royal lineage represents a good opportunity for understanding the configuration of the rural hinterland around the Bronze Age city, landscape changes in the late imperial era present major challenges, which could be resolved from the analysis of Corona and other remote sensing images. During the fifteenth century, the Ming government built a series of hydraulic projects around the cemetery to increase the water flow in the Grand Canal in order to help the cargo fleets 'climb' over the forty-meter elevation ridge, which is the highest point in the full length of the 1000 kilometers canal. A river was diverted to feed into the canal, while the flow was manipulated with several dozen shiplocks and three large reservoirs as 'water tanks'. The Bronze Age royal cemetery, therefore, was turned into an 'island' surrounding by a total of 175 km^2 of water surface from the large 'water tanks' that were used to control the canal flow (Fig. 5.3).

In 1855, a catastrophic flood from the Yellow River inundated the Grand Canal and filled all three water tanks with sediment. Today, the area has the deceptive appearance of a few tomb mounds over a flat plain with an abandoned canal meandering through it. An attempt to understand the Bronze Age rural landscape around the Lu royal cemetery, therefore, needs to account for the late imperial hydraulic projects that would have brought enormous changes to the distribution of archaeological sites, features, and artifacts amidst their construction, as well as the effect of the flood sediments that covered these reservoirs and their surrounding area. These reservoirs are so large that it could take several seasons of survey work to cover them (Fig. 5.4). In combination with historical maps of the hydraulic project (which are often not to scale), an analysis of the Corona and other remote sensing images for those extensively distributed dark sediments could help the archaeologists effectively address these landscape transformations in their research design and field practice (Shandong Provincial Institute of Archaeology et al. 2011).

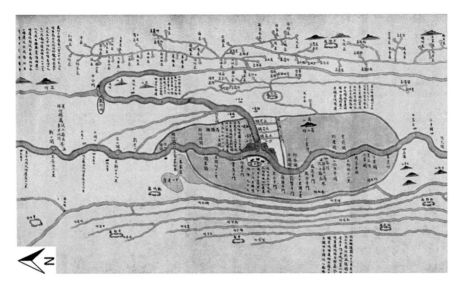

Fig. 5.3 The eighteenth-century historical map of the Nanwang hydraulic system from Niusheng yunhe quanyuan shuili qingxingtu (The hydraulic atlas for the canal transportation of nine provinces). Image source: Shandong Provincial Institute of Archaeology et al. 2011: 12

5.3.3   The Sacred Landscape

Like other places in the world, archaeological research in China is often focused on a specific phase in history, with a clear demarcation of the Neolithic, Bronze Age, and imperial periods. The information presented on published archaeological maps is selective, as it is based on the research orientation of the archaeological projects. Many archaeological maps only feature a distribution of archaeological sites from a single period, or a single type of sites. Clearly, this is not sufficient when the objective is to assess the recurrent patterns from a long time span in relationship to the landscape configuration. Corona images help create a visual juxtaposition of sites from different periods as a palimpsest so that issues of *longue durée* can be explored, e.g. routes of transportation and sites of religious pilgrimage, which were often associated with deep history.

The last example deal with the relationship between settlement sites and sacred landscape, which were often treated as different subject. Several mountains and low hills in and around the Qufu region were revered as sacred mountains in the early imperial period, which became major destinations for imperial pilgrimage in late first millennium B.C. (Li 2001, 2010; Wang 2011). These ritually significant places were regarded as centers of cosmological importance, even serving as *axis mundi*, the points of intersection between the natural and the supernatural world (Kleeman 1994). Royal tours to these sacred locations function as a symbolic exercise of control and represent the performative aspects of empire-building and world-making. These early imperial pilgrimages had a lasting influence on the ideology of kingship through Chinese history.

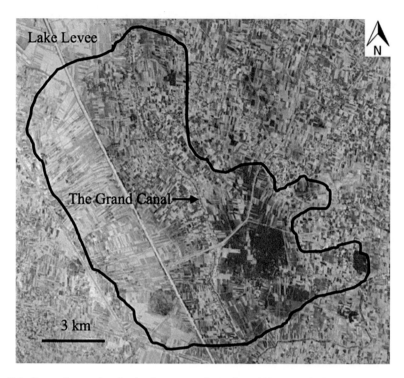

Fig. 5.4 Corona image showing landscape transformation of Nanwang hydraulic system around the Lu royal cemetery. Dark areas on the image represent the Ming water tanks inundated by the 1855 flood from the Yellow River. DS1115-2167DF039_39_d September 20, 1971 Longitude 116.36 Latitude 35.59

Corona images offer critical insight into the configuration of these sacred places in relation to archaeological features of different time periods and spatial scales. During the early imperial pilgrimage to Mt. Yishan, for instance, the First Emperor of Qin would have passed by the ruins of a Bronze Age city at the southern slope of this rocky hill. The Jiwangcheng site was once the capital of the Zhu state, a subsidiary of the Lu state during the middle first millennium BC. The close association of the archaic political center and the sacred landscape indicates that Mt. Yishan was probably a sacred mountain in the local ritual tradition well before the imperial conquest. An all-encompassing system of imperial sacred landscape, therefore, claims legitimacy through its incorporation of local history and memory.

In comparison to the intensive work done at Qufu, approximately 30 kilometers to the north, there is very little archaeological investigation at Mt. Yishan and the Jiwangcheng site. The correlations between archaeological features and image patterns established from analysis of known features at Qufu, therefore, help raise questions in this unknown region, which could guide field investigations. Preliminary observation of the Corona image reveals that Mt. Yishan figured prominently in the local political landscape long before the early imperial period (Fig. 5.5). From analyzing the Corona images of Mt. Yishan, we notice that the foundation platform of

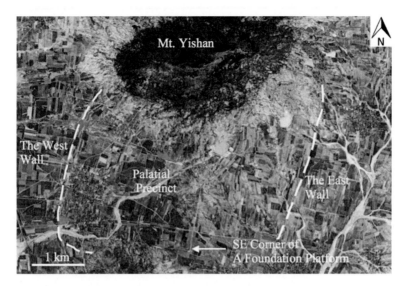

Fig. 5.5 The sacred landscape of Mt. Yishan (elevation 582 m) in relation to the site of Jiwangcheng, the capital city site of the Zhu state in the first millennium B.C. The wall of the Bronze Age city site at Jiwangcheng extended toward the slopes of Mt. Yishan in an effort to incorporate the sacred mountain as part of the urban landscape. DS1101-1069DF131_131_b September 20, 1967 Longitude 117.02 Latitude 35.31

the Bronze Age palace is aligned with the central peak. The eastern and western walls fan out toward the stony hill as if symbolically incorporating the sacred mountain into part of the urban configuration. The overall urban plan, therefore, evolved around the deeply entrenched notions of sacred landscape at Mt. Yishan, which eventually became part of the imperial sacred landscape.

5.4   Conclusion

Due to the restrictive access to high-resolution topographic maps in China, the relationship between major sites and their landscape is often underrepresented in archaeological publications. Corona images produce a highly effective visualization of a local landscape and a view of a lost landscape four decades ago that was often unavailable to field archaeologists themselves. This is even more critical when the off-sites and the surrounding terrain increasingly became the subject of inquiry in a landscape approach to archaeology. With its comprehensive coverage of Chinese landscape and open access, declassified Corona images could be used effectively to cope with these challenges. As research focuses shift from cemeteries and settlements to the archaeological landscape, Corona images will make a critical contribution to students and researchers interested in the history and archaeology of China.

Acknowledgements The field research in Qufu is a collaborative effort with the Shandong University, the Shandong Provincial Institute of Archaeology, and UCLA. It is supported by funding from the Henry Luce Foundation for East Asian Archaeology, the Ministry of Education of China 111 Project Grant, the Cotsen Institute of Archaeology, and the Faculty Research Grant at UCLA. The Office for Instructional Development at UCLA provided a teaching improvement grant to purchase the Corona images. The W. M. Keck Center for Digital Humanities offered its technological personnel to teach students of effective use of the imaging programs. I would like to thank Michael Harrower, Elaine Sullivan, Rachel Lee, Eric Fries, Fred Limp, Jackson Cothren, Jesse Casana, Luan Fengshi, Fang Hui, Wang Rui, Li Ling, Liu Jianguo, Zhang Li, Carrie Zhou, Stephanie Salwan, and Mandy Chan for their generous help.

References

Jing, Z., Rapp, G. Jr., and Gao, T. (1995). Holocene landscape evolution and its impact on the Neolithic and Bronze Age sites in the Shangqiu area, northern China. *Geoarchaeology 6*, 481–513.

Kleeman, T. F. (1994). Mountain deities in China: The domestication of the mountain god and the subjugation of the margins. *Journal of the American Oriental Society, 114*, 226–238.

Li, L. (2001). Qin Han cizhi tongkao (A study of Shrines and Ritual places in Qin-Han China). In L. Li (Ed.), *Zhongguo fangshu xukao (A study of the occult tradition in ancient China)* (pp. 187–203). Beijing: Dongfang Press.

Li, F. (2006). *Landscape and power in early China: The crisis and fall of the Western Zhou 1045–771 BC*. Cambridge: Cambridge University Press.

Li, L. (2010). Guren de shanchuan (Landscape of the ancient world). *Huaxia dili, 1*, 40–64.

National Bureau of Cultural Relics. (2007). *Zhongguo wenwu dituji (Archaeological atlas of China). Shandong volume*. Beijing: Zhongguo Atlas Press.

Rosen Arlene (2008). The Impact of Environmental change and Human Land Use on Alluvial Valleys on the Loess Plateau of China during the Mid-Holocene. *Geomorphology 101*: 298–307.

Shandong Provincial Institute of Archaeology. (1982). *Qufu Luguo gucheng (The ancient city site of Lu State in Qufu)*. Jinan: Qilu shushe.

Shandong Provincial Institute of Archaeology. (2000). *Zhongguo Linzi wenwu kaogu yaogan yingxiang tuji (The archaeological aerial photo-atlas of Linzi, China)*. Jinan: Shandong Atlas Press.

Shandong Provincial Institute of Archaeology, Chinese Academy of Cultural Heritage, Jining City Bureau of Culture, Wenshang County Bureau of Culture. (2011). *Wenshang Nanwang (Survey and excavation of the Nanwang hydraulic system in Wenshang)*. Beijing: Wenwu Press.

Ur, J. A. (2003). CORONA satellite photography and ancient road networks: A northern Mesopotamian case study. *Antiquity, 77*, 102–115.

Ur, J. A. (2006). Google Earth and archaeology. *The SAA Archaeological Record, 6*(3), 35–38.

Wang, R. (2011). *Bazhu jisi yanjiu (Research on the cult of the eight masters)*. Ph.D. dissertation. Department of Chinese Literature, Peking University.

Wilkinson, T. J. (2003). *Archaeological landscapes of the Near East*. Tucson: University of Arizona Press.

Part II
Multispectral and Hyperspectral Imagery

Chapter 6
Multispectral and Hyperspectral Technology and Archaeological Applications

Michael J. Abrams and Douglas C. Comer

Abstract This chapter describes the theory behind optical remote sensing using multispectral instruments. The fundamental principles include spectral reflectance, and spectral and spatial resolution. Wavelength regions of interest are the visible, reflected infrared, and thermal infrared. Both airborne and spaceborne sensors are described, with emphasis on those that provide free or low-cost data. The second part of the chapter presents an overview of the spectral remote sensing process, making mention of data interpretation and information extraction from multispectral data using standard image processing techniques. The chapters in the section to follow provide examples of the use of multispectral data for archaeological applications.

Keywords Multispectral • Spectral • Satellite and aerial remote sensing for archaeology • NASA • Image interpretation

6.1 The Universe of Multispectral and Hyperspectral Data

We outline here what you should know about multispectral and hyperspectral remote sensing before attempting to apply this technology to archaeological research and heritage management. Further information and in-depth treatments of remote sensing may be found in Campbell and Wynne (2011), Lillesand et al. (2007) and Jensen (2000). The technology and all applications of it deal with specific bands within the electromagnetic spectrum. Bands are typically identified by electromagnetic wavelength. Multispectral and hyperspectral bands include, but extend beyond, wavelengths that are sensed by the human eye and processed by the human brain. Depending upon the specific multispectral or hyperspectral device, these bands include those a bit shorter than visible light wavelengths, which are termed ultraviolet, and much more frequently, those having wavelengths somewhat longer than those in visible light, which fall into the near infrared (visible and near infrared, VNIR), reflected infrared (or short wavelength infrared, SWIR) and emitted infrared

D.C. Comer and M.J. Harrower, *Mapping Archaeological Landscapes from Space*,
SpringerBriefs in Archaeology, DOI 10.1007/978-1-4614-6074-9_6,
© Springer Science+Business Media New York 2013

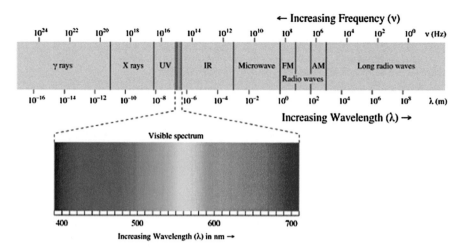

Fig. 6.1 Electromagnetic spectrum with visible light highlighted (Used under the terms of the GNU Free Documentation License on Wikipedia)

(thermal infrared, TIR) segments of the electromagnetic spectrum. The difference between multispectral and hyperspectral sensors is that the hyperspectral sensing process divides what is almost the same segment of the electromagnetic spectrum (made up largely of visible, near infrared, and infrared wavelengths) into many more, and therefore much narrower, bands. Hyperspectral sensors therefore produce data and images of greater *spectral resolution* than do multispectral sensors. Traditionally, hyperspectral refers to data acquired with continuous and contiguous bands of 10 nm or less in width in the VNIR and SWIR, and 15 nm or less in the TIR.

Multispectral and hyperspectral imagery are becoming more and more common in many fields, including engineering (including civil as well as disaster and emergency planning), health sciences, geography, agriculture science, history, commerce and business management (to name a few). Probably the most used and familiar multispectral imagery examples are natural color aircraft and satellite images used in such web applications as Google Earth, Google Maps, Mapquest, and Bing Maps, but since these programs only allow viewing imagery and do not provide access to the underlying data or digital number (DN) values of the images they cannot be used for many forms of analysis including many of those mentioned below.

Wavelengths longer than those in the infrared segment of the electromagnetic spectrum are microwaves, and are sensed by radar apparatus. Wavelengths shorter than the ultraviolet are so highly energetic that they generally pass through materials of interest to the archaeologist or are not reflected by them in ways that are informative. Wavelengths in the electromagnetic spectrum including those used in multispectral and hyperspectral remote sensing are seen in Fig. 6.1.

Sighted humans make use of visible light every day to make their way through the world safely and effectively. To understand how these and other bands can inform us about the environment around us, we will begin by considering how the environment is presented to humans by means of visible light. We will then consider

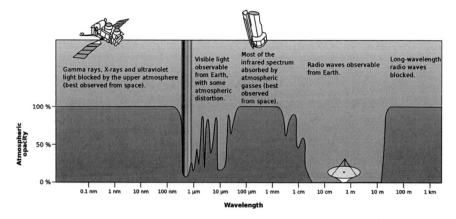

Fig. 6.2 Atmospheric interaction with different portions of the electromagnetic spectrum (Used under the terms of the GNU Free Documentation License on Wikipedia)

the processes and the technology by which all of the multispectral and hyperspectral bands are made available for analysis. Finally, with all of this in mind, we present an overview of ways that imagery generated from data provided by sensing technology has been processed to provide information that is pertinent to archaeological research and heritage management.

It is notable that visible light makes its way to earth with relatively little distortion by the atmosphere, compared to most other segments of the electromagnetic spectrum. This is seen in Fig. 6.2. Humans have probably evolved in ways that permit them to sense visible light for that reason and to match our visible spectrum to the wavelength region that corresponds to the maximum energy put out by our Sun. Reflected light produces the colors that humans see. Humans learn to associate colors with environmental characteristics that are pertinent to their safety and well-being. Most types of healthy vegetation are seen as green, for example, because the chlorophyll necessary to photosynthesis absorbs red and blue wavelengths, but reflects green. The sky on a clear day looks blue because the atmosphere scatters (reflects) blue wavelengths. Clouds appear dark because water vapor absorbs all bands of visible light. Taking visual cues occurs largely with little conscious thought; humans interpret the world around them by taking cues not only from tone or color, but also by texture, pattern, shape, shadow, size, and situation (context).

Just as the human eye senses reflected radiation from visible light, so do multispectral and hyperspectral devices sense reflected radiation from other portions of the electromagnetic spectrum. Figure 6.3 displays multispectral bands as they are sensed by the Landsat Enhanced Thematic Mapper satellite, a type of satellite in the Landsat series, which has probably provided more data and images than any other remote sensing platform. Color designations in the figure are somewhat arbitrary; for example, the "green" band from another sensor will typically span a slightly different range. The green band for the Geo-Eye 1 satellite is from 506 to 595 nm, in comparison, and a hyperspectral sensor would split the "green" segment of the electromagnetic spectrum into many smaller bands.

Landsat ETM+ Multispectral Bands			
Band #	Wavelength	Band Description	Common Applications
1	0.45 to 0.515	Blue	Distinguishing soil from vegetation, water penetration, deciduous vs. conifers
2	0.525 to 0.605	Green	Determining plant vigor (reflectance peak)
3	0.63 to 0.69	Red	Matches chlorophyll absorption-used for discriminating vegetation types
4	0.76 to 0.90	Near infrared	Biomass content, vegetative vigor
5	1.55 to 1.75	Shortwave infrared	Indicates moisture content of soil and vegetation, cloud/smoke penetration, vegetation mapping
6	10.40 to 12.50	Thermal infrared	Geological mapping, soil moisture, thermal pollution monitoring, ocean current studies
7	2.09 to 2.35	Shortwave infrared	Ratios of bands 5 & 7 used to map mineral deposits

Fig. 6.3 Bands as they are sensed by Landsat ETM+satellite

The "green" band can be used to determine plant vigor in the analysis of remote sensing imagery, just as humans intuitively use greenness to assess plant vigor in their daily lives; the stronger the green return, in general, the healthier the plant. Remote sensing technology, however, provides a way to quantify returns, and to combine them with returns from other bands in ways that can provide a more precise measure of vegetative health. For instance, Band 4, which is the near infrared band, can also provide an index of vegetative vigor. This is because healthy vegetation, although it absorbs red, strongly reflects near infrared. If this were not the case, heat generated by the absorption of near infrared radiation could damage plant cell tissue. Conversely, healthy vegetation absorbs red light, and so strong returns from the red band suggest that vegetation that depends upon chlorophyll for photosynthesis is not healthy. As we shall see, simple algorithms can be developed that synthesize data provided by returns from different bands in very useful ways.

6.2   Creation of Imagery

Sensed data collected by airborne or satellite platforms are generally provided in the form of images. Images are created by means of process that can be thought of as occurring in several steps, each of which can affect data quality. These steps are as follows:

6.2.1   Transmission of Electromagnetic Waves to Materials of Interest

For multispectral and hyperspectral remote sensing systems, the source of the radiation is the sun. Therefore, such systems are referred to as passive remote sensing systems. Other systems in which electromagnetic waves are transmitted from humanly constructed devices, such as radar and Lidar systems are known as active remote sensing systems. In both cases, angle of transmission affects the manner in which radiation returns from the material for reasons we describe later.

The nature of electromagnetic radiation remains mysterious in some ways: It is propagated by waves, but it behaves in some ways like particles. The particles are called *photons* or *quanta*.

Wave theory tells us the velocity of light equals wavelength times wave frequency:

$$\mathbf{c} = \mathbf{v}\lambda$$

Particle theory informs us that the energy of a quantum is inversely proportional to its wavelength:

$$\mathbf{Q} = \mathbf{hc}\,/\,\lambda.$$

Thus:
The longer the wavelength, the lower its energy content, and
Wavelength and wave frequency vary inversely.

Together, these carry with them some important implications:

- Devices that sense longer wavelengths must "view" larger areas of the earth in order to detect energy signal sufficient to produce images.
- That being so, the spatial resolution of images generated from sensed longer wavelengths is generally lower because the pixels from which the image is made must be larger.
- At the same time, shorter wavelengths are more absorbed and scattered by the atmosphere
- Therefore, sensed radiance of shorter wavelengths varies more with atmospheric conditions.

6.2.2 *Interaction of Radiation with the Atmosphere*

Radiation is effected by passing through the atmosphere as it travels toward earth and, after being reflected by materials there, as it returns to the airborne or satellite sensing device. Energy is scattered and absorbed by the gases and particles in the atmosphere. Areas of the electromagnetic spectrum not greatly influenced by absorption are called atmospheric windows as seen in Fig. 6.2. These windows depend upon the varying energy of different waves and how each reacts as it is propagated. Atmospheric conditions are for our purposes "random," which complicates interpretation of images. Depending upon how images are to be interpreted, it might therefore be necessary to "correct" for atmospheric conditions, although the random nature of atmospheric interference makes this difficult. Because the

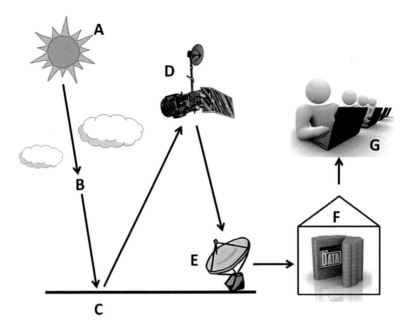

Fig. 6.4 Schematic diagram of transmission of energy from the sun, reflection from the surface, interaction with the intervening atmosphere, and capture by sensors. The data are converted to electronic signals, and sent to a ground receptor

atmosphere affects longer wavelengths less, it might be more advantageous to analyze images generated from them, even though such images might be of lesser resolution. For example, blue and violet are strongly scattered in ways that vary according to atmospheric conditions. Red and near infrared are often more reliable because they are less effected. Many useful algorithms have therefore been developed for these bands.

6.2.3 Interaction with the Target

A schematic energy path is shown in Fig. 6.4. Visible and reflected infrared energy from the Sun (A) passes through the atmosphere (B) where it is scattered and absorbed before reflecting from the surface (C). The reflected light passes back through the atmosphere, and is sensed by the airborne or spaceborne instrument (D), where it is converted from photons to an electrical voltage by light sensitive detectors. For thermal infrared data, additional source of sensed energy comes from the surface itself radiating energy in various wavelengths as a function of the surface's temperature and emissivity. Again the atmosphere contributes to the sensed signal by modifying the incoming and outgoing thermal energy, and contributing to it through radiation of the atmosphere itself.

6.2.4 Recording of Energy by the Sensor

After the energy has been reflected (bounced, scattered, etc.) by the target, we require a sensor to collect and record the electromagnetic radiation (Fig. 6.4). If that sensor is not in contact with the target, it is remote – in our case, very remote, because it is carried by an aircraft or a satellite. Sensor malfunction can be a factor. Sensors and the platforms that carry them are highly complex, and when components malfunction, it is usually impractical to replace or repair them. In time, every instrument wears out. Also, there are engineering challenges for which perfect solutions have yet to be devised. For example, landscapes that change rapidly from bright to dark or vice-versa can produce "memory error." These problems create anomalies in data collection and imagery created from data. The good news is that they are non-random, and so can often be corrected to some extent.

Returns are transformed by the sensing device into numbers, more specifically, "digital numbers" (DNs). These are assigned to pixels, which form images, using a coordinate system that models the landscape for which returns have been obtained. In our Landsat example, the spatial precision of this model is high (about 15–30 m), and in the absence of a major system malfunction, reliable. Other satellites accomplish this with lesser degrees of precision. Detectors and data systems are designed to produce a linear response to incident spectral radiance, that is, incremental increases in spectral radiance are converted to DNs by a linear function. DNs that make up a digital image are recorded over numerical ranges such as 0–255, 0–511, 0–1023, 0–2046. These are dictated by the set of integers produced by 8-, 9-, 10, and 11-bit computer coding scales (i.e., $2^8 = 256$, $2^9 = 512$, $2^{10} = 1,024$, $2^{11} = 2,016$). Such scales are different for different sensors: Landsat TM is 8-bit, for example, while the old Landsat Multispectral Scanner (MSS) has a 6-bit sensor (and so only 64 gray levels).

6.2.5 Transmission, Reception, and Processing

The energy recorded by the sensor is transmitted, usually in electronic form, to a receiving and processing station where the data are processed into an image. Before the image is provided to the user, DNs can be further transformed, which is a way to generalize from hundreds or thousands of pixel values to fewer. Most frequently, values from 0 to 255 are assigned to each pixel. Assigning values 0–255 to pixels making up a gray scale image or 0–255 for each of the color channels (red=R, G=green, B=blue) will produce a color image that satisfies most needs. Figure 6.5 illustrates the concept of RGB composites and the resulting combination of colors represented by DN triplets. For example RGB=0,0,0 represents black, and RGB=255,255,255 produces white in this additive system. There are many other color schemes for producing color composites from three different bands, such as CMYK. The RGB world for creating color satellite images probably evolved from

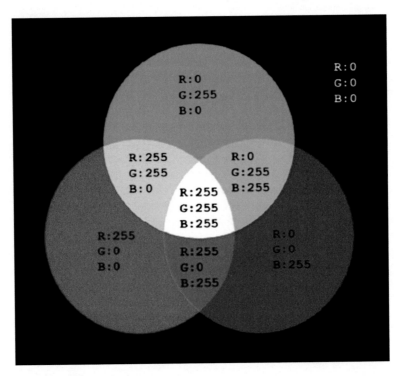

Fig. 6.5 Venn diagram representing the concept behind RGB color space in a digital environment using the most common industry standard of 256 shades per color channel (After Hixson)

television monitors and their use of RGB guns to produce color variation, as early image processing hardware systems featured TV monitors for displays.

6.2.6 Concepts of Resolution

There are four types of resolution when discussing satellite imagery in remote sensing. Spatial resolution is used to refer to the instantaneous field of view (IFOV) or pixel size of the instrument capturing image data. Strictly speaking, resolution is twice the IFOV, but the term has been loosely used interchangeably with IFOV. Current commercial civilian satellites provide multispectral data with pixel sizes as small as 40 cm (such as GeoEYE); government operated satellites more typically deliver data with pixel sizes of 10–30 m (Fig. 6.6). Spectral resolution refers to the wavelength interval size that the sensor is measuring. Multispectral scanners in the reflected part of the spectrum typically have relatively broad bands of 50–100 nm; hyperspectral scanners may have bands of 5–10 nm in width. Temporal resolution refers to the number of days that pass between successive image acquisitions for the same place on the earth surface. Landsat satellites have a repeat period of 16 days, for example; WorldView-1 from Digital Globe has capability of repeat coverage of

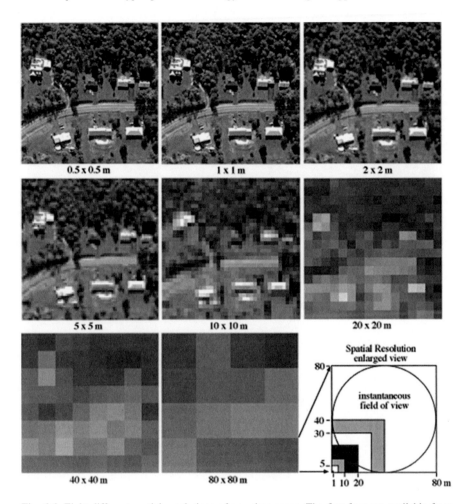

Fig. 6.6 Eight different spatial resolutions of an urban scene. The first four are available from commercial companies, and the last four are typical of civilian government data providers (After Jensen 2000)

1.7 days. Lastly, radiometric resolution is defined as the ability of a sensor to record many levels of brightness. It is the bit-depth of the sensor (number of grayscale levels) and is expressed as 8-bit (0–255) up to 16-bit (0–65,535).

6.2.7   *Analysis and Interpretation*

Archaeologists will in most cases receive remotely sensed data in a pre-processed image. Simply looking at the image in the form it was provided can, of course, be informative. An aerial or satellite image provides a perspective very different from

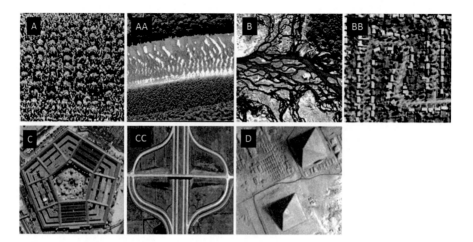

Fig. 6.7 (**A**) and (**AA**): Texture in images interpretable as trees, and sandy river channel; (**B**) and (**BB**): Pattern in images interpretable as meandering river channels and suburban housing project; (**C**) and (**CC**) Shape in images interpretable as the Pentagon building and a freeway interchange; (**D**) Shadow in images interpretable as pyramids (After Jensen 2000)

that which can be obtained on the ground. Important spatial relationships among features of interest can be observed from this synoptic point of view, patterns and relationships emerge that might have gone unnoticed if one had encountered the features that form them one or a few at a time on the ground. In identifying these relationships, the human eye and mind can draw upon the observation of texture, pattern, shape, shadow, size, and situation (context), just as it does in daily life (Fig. 6.7). These patterns and relationships would be seen if they were mapped, of course, but in the absence of a synoptic view of them, one might not suspect that they should be mapped. Identified spatial relationships can be quantified and analyzed. Often, doing this will suggest returning to the field to better record and understand the features of interest. *No analysis is reliable in the absence of information obtained on the ground; the analysis of imagery is not a substitute for fieldwork.* Image analysis and fieldwork are complementary, an archaeologist must move back and forth between the approaches to obtain reliable information and to develop a useful interpretation of material observed in both ways.

By now, almost all archaeologists have made use of satellite images that can be viewed from personal computers, and a good bit of the types of analyses and interpretation described in the paragraph above can be conducted simply by availing oneself of these images. The images will be in color or black-and-white and of varying degrees of spatial resolution, although high-resolution images are increasingly available.

Taking more complete advantage of multispectral and hyperspectral data, however, can only begin after acquiring images that contain not just bands of visible light, but also those in the near-infrared and infrared ranges, and software that can manipulate these bands. Available imagery and software are increasingly varied, and a search on the Internet will soon yield many examples of both.

In general, the type of multispectral and hyperspectral images an archaeologist might choose for her or his purposes will depend upon the objectives of a specific research or heritage management effort and available funding. If knowledge of the environment within which the area of study is located is a desired outcome, there is a good deal of relatively low-spatial resolution imagery available at no or low cost. The Landsat imagery described above is perhaps paramount among them. Landsat imagery is free, and provides a history of environmental change over the past several decades. Early Landsat images can provide environmental information that is more pertinent to understanding ancient occupations, because it might have been obtained before agricultural, industrial, or residential development. Other NASA imagery includes that obtained by data collected by the ASTER (Advanced Spaceborne Thermal Emission and Reflection Radiometer) multispectral satellite and the Hyperion hyperspectral satellite, which can be obtained at relatively low cost.

Higher spatial resolution multispectral and hyperspectral imagery at the time of this writing can be obtained through private-sector vendors, but is costly. Among the best-known vendors of multispectral high-resolution imagery are Geo-Eye and Digital Globe. The cost is less for images that have already been obtained, and quite a bit more for custom collection of data and development of images from that data. Images are sold at a cost that also differs according to the precision of the imagery, the more spatially precise, the more expensive it will be to acquire.

With this in mind, it is wise to think carefully about the objectives of the specific research or preservation project. Many archaeologists desire high-resolution imagery, in which they can discern relatively small features. While this is so, environmental understanding does not always require such a fine-grained approach. If one is searching for individual archaeological sites and features that cannot be seen in the color images that can readily be obtained at no cost on the Internet, one must understand that she or he is entering a realm in which more sophisticated knowledge of image enhancement and analysis is required. This more sophistic knowledge, of course, can also be of use in acquiring the most useful information possible from lower resolution images.

More sophisticated enhancement of or analysis of the images depends, briefly, on reassigning numbers to the pixels that make up an image. This can be done in both uninformed and informed ways. Most image analysis software and some geographical information system software (GIS) provide drop-down menus that can be used to enhance and analyze images. Almost invariably, archaeologists who obtain both imagery and software begin by experimenting with drop down capabilities. Among the most commonly used of these is contrast stretching. Values are seldom evenly distributed across all pixels. If one is viewing an image composed of pixels having values from 0 to 225 (which is quite standard) and for which values cluster largely within the range of 60–160, one might be able to discern features or areas of interest better by "stretching" those clustered values so that they are assigned to a greater number of pixels. The drop down provides several ways in which this can be done. Even with a minimal understanding of how this transformation is occurring, the results are sometimes informative; one can in fact see an area or pattern that is informative by its presence. Others include simply reassigning bands to the red, green, and blue channels that make up an image seen on a computer screen, which is easily

Fig. 6.8 The Temple of Amarna, Egypt as seen on Google Earth imagery (*left*). The right half of the image has had contrast stretch applied to bring out details related to man-made structures

done with most image analysis and GIS software, or segmenting an image into a number of areas that can be specified based on similar pixel values. This can be done before or after contract stretching.

Informed image enhancement and analysis requires training or study. Much of the informed work falls within one of these four broad categories of computer-assisted operations. An informed user of the technology should develop some proficiency with each:

- *Image rectification and restoration*
  Most remote sensing imagery is obtained in a processed format. However, sometimes further treatment is necessary to remove image distortion, orthorectify the data, radiometrically correct the data (i.e., discerning original radiance values), and remove noise.
- *Image enhancement*
  Various simple techniques fall into this category: contrast manipulation, spatial feature manipulation (filtering), and multi-image manipulation. Figure 6.8 is an illustration taken from Google Earth data of the Temple of Amarna, Egypt. The left half shows the original image; the right half has had a contrast enhancement applied, to exaggerate the contrast and make building features easier to recognize. This is an example of how basic image processing software can be used in a relatively uninformed way using publicly available, high spatial resolution data.

- *Image Classification*

  Image classification techniques have the overall goal to automatically categorize all pixels in a scene into unknown or known classes. The former process is called unsupervised classification, and assigns pixels to groups based on similarity of their spectral characteristics. Supervised classification requires prior knowledge of the considered classes and significant user interaction at many steps in the process. The various supervised classification algorithms all compare the spectrum of an image pixel to a known spectrum (laboratory- or field-measured, or extracted from the image itself from a known class) and assign to classes using statistical measures. An example of supervised classification is by Menze and Ur (2012), and involved using hundreds of ASTER multitemporal images, Corona images, and topography data to detect anthrosols and map the pattern of human settlements in Mesopotamia (Fig. 6.9). They claim to have mapped more than 14,000 settlement sites, site volumes, and exchange networks between sites.

- *Data merging*

  Altaweel (2005) reported an investigation using Corona and ASTER images to look for hollow ways at Tel Kashaf near Nineveh. These are shallow linear depressions connecting ancient sites. They have different soil, moisture and vegetation characteristics than surrounding terrain. On the Google Earth image (Fig. 6.10a) little can be seen. The high spatial resolution Corona image (Fig. 6.10b) shows linear features coming from Tel Kashaf that are interpreted as hollow ways; the ASTER thermal image shows other linear features due to temperature contrasts between packed earth and adjacent field materials (Fig. 6.10c).

An archaeologist who wishes to use multispectral and hyperspectral technology in an informed way should master these basic skills as much as possible.

6.2.8   Application

Multispectral and hyperspectral remote sensing have long been used to investigate and characterize what is generally thought as the natural environment (Johnson, 2006, Lasaponara and Masini, 2012; Parcak, 2009; Wiseman and El-Baz, 2007). Archaeologists typically understand that the natural environment as a place undisturbed by human activity long ago ceased to exist. Humans and humans ancestors with some level of cultural development have inhabited the Old World for hundreds of thousands, and even millions, of years. The New World has been populated by humans for perhaps 15,000 years, but in that time they have reach every part of the hemisphere. The environments that they encountered were an enormous influence on settlement patterning, and ultimately on the wide variety of cultures they developed and the societies shaped by those cultures. As time went on, humans altered the environment in increasingly noticeable ways. It is this interplay of environment and culture about which multispectral and hyperspectral remote sensing can most inform us.

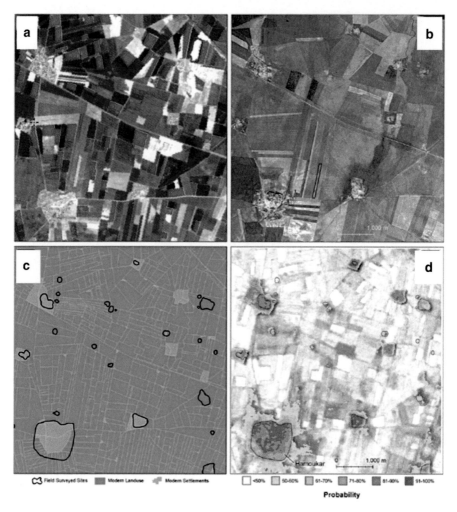

Fig. 6.9 (a) ASTER image for the Hamokur region in false color infrared; (b) Corona image; (c) Sites, modern villages and contemporary land use; (d) Anthrosol probability derived from multitemporal ASTER analysis (After Menze and Ur 2012)

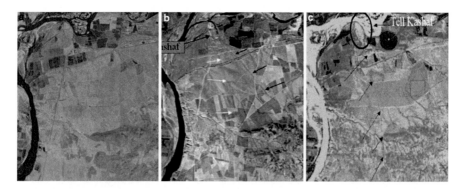

Fig. 6.10 (a) Google Earth image of Tel Karshaf near Nineveh; (b) Corona image showing linear features interpreted as hollow ways; (c) ASTER thermal infrared image showing additional linear features (After Altaweel 2005)

Until recently, the use of aerial and satellite remote sensing by archaeologists had received attention for the most part because archaeological sites had been discovered. The environmental perturbations associated with human settlement can indeed help us to find sites. In the end, however, archaeologists will contribute to our knowledge of human history not so much because they are finding new sites with multispectral and hyperspectral imagery as because the analysis of it can provide evidence pertinent to why they are there and when and how this happened.

Acknowledgments Work done by Abrams was performed at the Jet Propulsion Laboratory/ California Institute of Technology under contract with the National Aeronautics and Space Administration.

Bibliography

Altaweel, M. (2005). The use of ASTER satellite imagery in archaeological contexts. *Archaeological Prospection, 12*, 151–166.

Campbell, J., & Wynne, R. (2011). *Introduction to remote sensing*. New York: Guilford Publications.

Jensen, J. (2000). *Remote sensing of the environment: An earth resource perspective*. New York: Prentice Hall.

Johnson, J. (2006). *Remote sensing in archaeology*. Birmingham: University of Alabama.

Lasaponara, R., & Masini, N. (2012). *Satellite remote sensing: A new tool for archaeology*. New York: Springer.

Lillesand, T., Kiefer, R., & Chipman, J. (2007). *Remote sensing and image interpretation*. New York: Wiley.

Menze, B., & Ur, J. (2012). Mapping patterns of long-term settlement in Northern Mesopotamia at a large scale. *PNAS, 109*(14), E778–E787. www.pnas.org/cgi/doi/10.1073/pnas.1115472109.

Parcak, S. (2009). *Satellite remote sensing for archaeology*. New York: Routledge.

Wiseman, J., & El-Baz, F. (2007). *Remote sensing in archaeology*. New York: Springer.

Chapter 7
Petra and the Paradox of a Great City Built by Nomads: An Explanation Suggested by Satellite Imagery

Douglas C. Comer

Abstract Petra has been seen by many as an historical and archaeological anomaly, a city with all the urban amenities seen in other great cities of the ancient world that was built by a famously nomadic Arab society. In the centuries before the advent of the Common Era, the Nabataeans amassed great wealth by controlling the trade in precious commodities, especially spices and incense, over the Arabian Peninsula. Written accounts from the fourth century B.C. indicate that so committed were the Nabataeans to a nomadic lifestyle that agriculture and the construction of houses were outlawed (Starky, 1955: 87). Yet by the first two centuries A.D., as Rome tightened control over trade in the region, many magnificent tombs and civic structures had appeared in Petra, and during this time there was a proliferation of village and temple complexes in Nabataea. Enhancement of Landsat satellite imagery suggests that villages and civic architecture appeared only in those areas where soils were arable and the terrain was amenable to the diversion of sparse rainfall to those arable soils. The appearance of Nabataean villages, temples, densely settled urban areas, and the famous tombs and monuments of Petra might therefore have emerged together as facets of a way of life appropriate to an agriculturally based economy that did not exist in Nabataea before that time.

Keywords Nabataea • Petra • Landsat • Caravan trade • Levant

7.1 Introduction

In 1816, the Swiss adventurer Johanne Burkhardt, in the guise of a Muslim pilgrim searching for the tomb of Aaron, made his way to Petra, which many had come to believe existed only in imaginative tales handed down over centuries, like Atlantis or Shangri-La. In what is now southern Jordan, he found spectacular tombs cut into the colorful walls of sandstone canyons, surrounding the ruins of a large city. Decades of archaeological research have since established that by the first century A.D., Petra

D.C. Comer and M.J. Harrower, *Mapping Archaeological Landscapes from Space*,
SpringerBriefs in Archaeology, DOI 10.1007/978-1-4614-6074-9_7,
© Springer Science+Business Media New York 2013

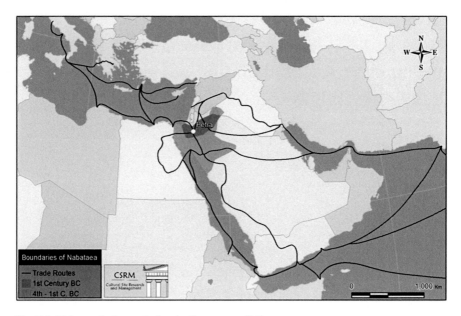

Fig. 7.1 Nabataea before and after the first century B.C.

was filled with temples, houses, and public spaces, including theaters, a *nyphaeum*, and a *paradeisos*, complete with pond and artificial island. Temple-village-agricultural field complexes had by then sprung up between Petra and the Dead Sea and in the lands west of the Wadi Araba that led to the Roman port of Gaza. Roman roads ran through Nabataea, and ships plied the waters surrounding the Arabian Peninsula.

Yet Petra was the capital of Nabataea (Fig. 7.1). For hundreds of years before the time of Christ, the Nabataeans had amassed enormous wealth by pursuing a nomadic lifestyle, one that pre-adapted them to the control of trade routes across the vast interior desert of the Arabian Peninsula (Starky 1955: 84). By camel caravan, they moved frankincense, myrrh, spices, gold, silver, pearls, silk, and other precious commodities from Arabia, South and Southeast Asia, and Africa to ports on the eastern coastline of the Mediterranean (Graf and Sidebotham 2003: 71–72). The Arabian Peninsula itself was their home. Only they knew the sources of water and hiding places there. Others who might try to convey these precious trade goods through the Red Sea found it, infested with pirates, and often these were Nabataeans pirates.

Nabataean society from the fourth through the first centuries B.C. (1996: 56) can be inferred from the account provided by Diodorus Siculus in his Universal History (2.48-49; 19.94-100). This was probably informed by Hieronymous, who was with the army commanded by one of Alexander's generals, the renowned Antigonus Monopthalmus, who attacked Petra twice in 312 B.C. (Starky, 1955: 84–85). Diodorus Siculus tells us that the Nabataeans were very wealthy, but entirely nomadic. They had knowledge and skills allowing them to build cisterns in the desert to supplement the very few natural sources of water. When attacked, the Nabataeans took their flocks and possessions to the deserts of the Arabian

Peninsula, where only they could survive. Diodorus Siculus also tells us that houses, fields, temples, or anthropomorphic images were prohibited by Nabataean law. Their mobility was clearly their greatest and practically unassailable defense at this time.

7.1.1 Agriculture

Yet nomads had not always occupied Petra and the land of the Nabataeans. The famous site of Beidha, near the Petra city center, has both Natufian and Pre-Pottery Neolithic B (PPNB) occupations. The Natufians at Beidha were incipient agriculturalists in the eleventh century B.C.; the PPNB occupation was by agriculturalists and herders, who built masonry structures between 7,200 and 6,500 B.C. There are numerous other Pre-Pottery Neolithic sites near Petra, and within the landscape that would be Nabataea, many more, including both Pre-Pottery Neolithic A and B. They are seen at Jericho, for example, dating from 8,500 to 6,000 B.C. The Edomites, who practiced agriculture, occupied the area in which Petra was constructed in the seventh and eighth centuries B.C. Evanari et al. (1971: 17) report the discovery of an agricultural farmstead dating to approximately 1,600 B.C. in the Negev.

Agriculture had been practiced for millenia in this area until it was depopulated sometime around the sixth or seventh centuries B.C. It is the period from approximately the sixth to the third century B.C. when no settlements are found in the Negev. In the third century, settlements reappear here; these are Nabataean (Evanari et al. 1971: 18), and seem to have been established to protect trade routes between Petra and the Mediterranean. The period of nomadism thus corresponds to that of the rise of Nabataean control over trade. No Nabataean pottery has been recovered that antedates the first century B.C. (Negev 1986: 29). As Negev says, the Nabataeans, like other nomads, had no need of pottery. The reappearance of Nabatean pottery is very likely linked to the gradual reappearance of agriculture in the region.

The analysis of satellite imagery suggests strongly that the Nabataeans began to engage in agriculture just as the grip they held on trade routes was loosened, and embraced it more fully as Rome established its control by increments over the Nabataean landscape and the trade routes there. As Benz has argued concerning the socio-ideological changes that accompanied the Neolithic, the perception of the landscape differs between foragers and farmers. Group identities are intensified by the construction of communal structures and the rituals that take place there; this is essential to establishing the new social rules that accompany the sedentary village life (Benz 2010: 77). Temples and villages sprang up in Nabataea in the first and early second century A.D., but only in places where the landscape was conducive to agriculture.

7.1.2   Attenuation of Nabataean Control Over Trade

Pompey waged his famous war on the Mediterranean pirates in 66 B.C. From that time onward, Rome gradually made maritime trade safer throughout the region, including the Red Sea, cutting into Nabataean profits. Rome also moved to take control of overland trade routes. Augustus instructed the prefect of Egypt, Aelius Gallus, to discover them. The Nabataeans expressed their willingness to cooperate with Aelius Gallus, but in 25 B.C., Syllaeus, a Nabataean minister, led him on confusing paths, which the Romans did not think were the routes that had been held secret by the Nabataeans for so long.

Jean Starky has described what happened next (1955: 94):

> The failure of the expedition of 25 B.C. and the bad will of the Nabataeans led Augustus to siphon off the major part of the trade to Egypt, annexed to Rome by 30 B.C. This was the accepted route by A.D. 18, for Strabo, who had just finished his *Geographica*, specifically notes that most of the merchandise was then discharged at Myos Hormos, a port a bit to the north of Leucos Limen, then transported by camel to Coptos, and then down the Nile to Alexandria. In the time of Malichus II (A.D. 40–70), a Nabataean agent resided at Leuce Come and collected a 25% tax on imports. This information from the author of the Periplus suggests that a percentage of the same order was levied a half century earlier and gives some idea of the riches that were accumulated in [the Nabataean city of] Petra. In his Natural History, Pliny the Elder alludes to innumerable shipping charges, taxes and tips which burdened the goods between Arabia Felix and Gaza, and which could finally increase the price of items a hundredfold.

Those vying for political leadership in Rome coveted this wealth. Power depended upon equipping, paying, and supplying benefits to the military, and providing sustenance and diversions to the general populace of Rome. The latter was difficult because of a grain shortage that had its roots in the Second Punic War, 218–201 B.C., when farmers destroyed their lands in order to deprive the Carthaginian army of provisions. The Roman elite subsequently bought up these farms and consolidated them to larger and consequently more productive landholdings called *latifundia*. *Latifundia* owners typically made great profit because they used cheap labor, often slaves captured during Roman wars.

If the new system restored a degree of agricultural production in the Italian Peninsula, it was insufficient to provide sustenance to the ever-larger numbers of displaced peasants who flocked to Rome. During the Third Punic War (149–146 B.C.), Carthage was destroyed. With the defeat of Marc Antony and Cleopatra at the Battle of Actium in 31 B.C., Roman hegemony of the Mediterranean was complete. Octavius (formerly Octavian, soon to be Caesar Augustus) moved rapidly to implement the Roman style of management in the lands to the east of the Mediterranean and in North Africa. Towns designed in the Roman style as well as roads, forts, temples, and *latifundia* that were needed to produce grain for Rome became common.

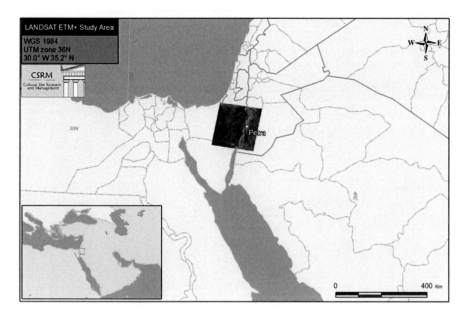

Fig. 7.2 Study area, in which images produced by Landsat ETM+ satellite bands were analyzed

7.1.3 The Landscape in Aerial and Satellite Imagery

Nelson Glueck conducted a pedestrian survey of the region to the north of Petra in the 1930s. In 2007, what appeared to be field notes for the survey were found by Joseph Greene of the Harvard Semitic Museum at the archives there. Examination of these documents by the author and an intern from Carnegie Mellon, Mary Grace Joseph, revealed that they contained not only notes from the regional survey (a novel approach to archaeological fieldwork at the time), but also notes from Glueck's excavations at the Nabatean High Place of Sacrifice called Khirbet et-Tannur (Glueck, 1937), an associated village, Khirbet edh-Dharih, and several other temples and Nabataean villages in an area between Petra and the Dead Sea. This area is seen in Fig. 7.2.

Aerial photos that Glueck obtained through the RAF in the 1930s were included in his notes. He interpreted them to be of Nabataean field systems, which he examined on the ground (Fig. 7.3). In a 1937 article about Khirbet et-Tannur (Glueck, 1937), Glueck noted that the Nabataeans had cultivated every bit of soil in Wadi Hasa, the valley in which Tannur is located, and that they had built terraces to hold soil in place.

The landscape of most of Jordan is an arid one. The Jordan River flows through Wadi Araba, part of the Great Rift, which extends to the south, far into the African Continent. As one moves to the east from Wadi Araba, the terrain rises steeply, and precipitation quickly decreases.

Fig. 7.3 Photo obtained in 1928 by the British Royal Air Force, in which Nelson Glueck detected Nabataean fields

Only 20% of Jordan receives more than 200 mm of precipitation each year, which is considered the minimum for rain-fed agriculture, and by a number of estimates only 6–7% of land in Jordan is arable (Metz 1989: 167). In the area we consider here, between Petra and the Dead Sea, rainfall varies between 100 and 300 mm per year, and large rainfall fluctuations occur from year to year. A sustainable agriculture therefore depends upon a landscape of the sort for which Petra (Greek for *rock*) is well known: one with large areas of rock or hard, sloping ground that sheds water, which then collects in flat areas downslope. The water is concentrated in these areas, so that the volume received there is much greater than would be the case were the landscape one of gently sloping soils. Terraces such as Glueck mentions are a way to provide locations at which water and the soils carried by it collect, rendering these places arable when they otherwise would not be.

Following Glueck's lead, we plotted the locations of the Nabataean fields, temples, and villages that he found, as well as others discovered in later years, in a GIS project.

Table 7.1 Electromagnetic bands sensed by NASA Landsat ETM + satellite

Band	Spectral range (microns)	Electromagnetic spectrum	Ground resolution (m)
1	0.45–0.515	Visible blue-green	30
2	0.525–0 .605	Visible green (reflected)	30
3	0.63–0.690	Visible red (reflected)	30
4	0.75–0.90	Near infrared (reflected)	30
5	1.55–1.75	Mid-infrared (reflected)	60
6	10.40–12.5	Thermal infrared (emitted)	30
7	2.09–2.35	Mid-infrared (reflected)	30
Pan	0.52–0.90	Visible light (reflected)	15

An analysis of NASA Landsat imagery was then conducted to determine aspects of the environment that might have been associated with the appearance of these fields, temples, and villages, which occurred in most cases in the late first century and early second century A.D., just before and as Nabataea was annexed by Rome in 106 A.D.

The NASA Landsat program has launched a series of satellites beginning in 1972. Landsat satellites record energy reflected and emitted from Earth in several bands of the electromagnetic spectrum. The author conducted a landscape analysis that was suggested by Glueck's notes, using images developed from Landsat ETM+ data. Bands sensed by the Landsat ETM + satellite are seen in Table 7.1. The image that we analyzed was acquired on March 8, 2002. Prior to the analyses described below, we conducted radiance and atmospheric corrections on the Landsat image.

To detect locations of arable soils, a normalized difference vegetative index (NDVI) was calculated. This index is commonly used to gauge the health and vigor of vegetation. It is generated by the formula NDVI = (NIR-RED/NIR + RED), where NIR is near infrared radiation and RED is light in the red wavelength. Healthy vegetation strongly reflects near infrared radiation, which is not very useful for photosynthesis, but can heat and cause damage to plant tissue. Healthy vegetation, typically reflecting green light, absorbs red light. When vegetation is less healthy, it usually is less green and reflects more red light. The formula works by linking the phenomenology of these two bands of the electromagnetic spectrum.

In Fig. 7.4, arable soils are colored red, which indicates the presence of vegetation today as determined by the results of the NDVI. This healthy vegetation is evidence of arable soils, rare in Jordan. Yellow dots indicate terraced fields or such fields found in association with farmsteads or villages, blue dots represent locations of temples.

A Tasseled Cap Transformation of several Landsat ETM + bands was performed in order to highlight other environmental attributes that made agriculture possible in the study area, e.g., the presence of rock or hard slopes that concentrate precipitation into flat areas, which are sometimes manmade terraces. The transformation was developed by Kauth and Thomas (1976), and converts several image bands into new ones, which provide indices of *brightness*, *greenness*, and *wetness*. In the arid landscape of the study area, these indices are informative. In Fig. 7.5, the intensity

Fig. 7.4 Image generated by a normalized difference vegetation index (NDVI) using Landsat ETM + satellite bands. Arable soils are shown in *red*. Note relationship of *yellow dots* indicating terraced fields or such fields found in association with farmsteads or villages and *blue dots* represent locations of temples with area of arable soils

of the color red indicates degree of wetness; of green, greenness, or presence of vegetation; and blue, brightness, or rock or hard slopes. Both the yellow and blue dots can be seen in areas of the landscape that combine these attributes.

Figure 7.6 is an oblique view of a temple, Ramm, obtained using Google Earth, which illustrates very well the typical spatial relationship of temples to the

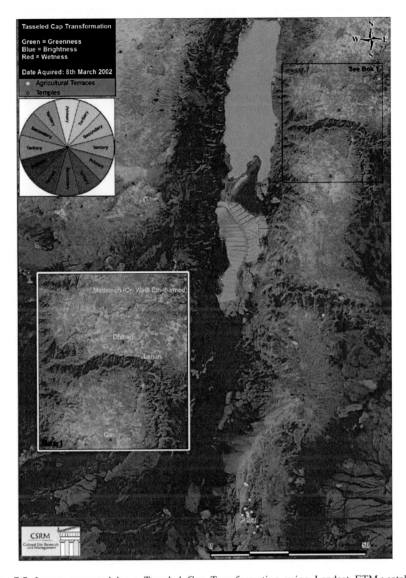

Fig. 7.5 Image generated by a Tasseled Cap Transformation using Landsat ETM + satellite bands: *blue* = brightness, *green* = greenness, *red* = wetness (Note again locations of *yellow and blue dots*)

environmental characteristics highlighted by the NDVI and Tasseled Cap analyses. Note that the landscape in which the temple is located is predominantly rock. Temples are located adjacent to arable soils (seen as red in the NDVI image and as yellow or orange in the Tasseled Cap), and near or sometimes on rock prominences, seen as blue in Tasseled Cap images.

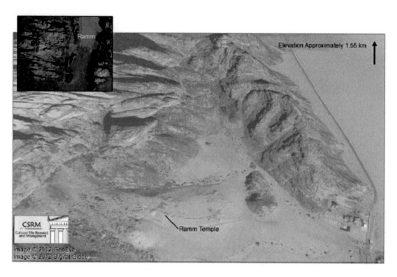

Fig. 7.6 Oblique view of Nabataean temple (Ramm) in Google Earth. Inset is an NDVI image, showing temple in spatial relation to arable soils (seen as *red* in NDVI) and rock slopes that channel water to arable soils, making agriculture possible

Analyzing the landscape by the use of readily available satellite imagery provides us with information that is pertinent to the endless dialectic between the environment and the uses to which humans put it. In this case, the correspondence between the chronologies established by the analysis of subsurface findings, such as that offered by Negev, and the dates at which field, villages, and temples appear in locations highly suitable for agriculture is striking. It suggests that the Nabataeans constructed the monumental architecture seen at Petra, Khirbet et-Tannur, Dharih, and other Nabataean settlements during the time when the Nabataeans were embracing the practice of agriculture in earnest, and not before, when they were culturally committed to a nomadic way of life.

References

Benz, M. (2010). Changing landscapes: Changing societies? An anthropological perspective. In B. Finlayson & G. Warren (Eds.), *Landscapes in transition* (pp. 77–85). Oxford: Oxbow Books.

Diodorus Siculus. Universal History (2.48-49; 19.94-100). English edition: Oldfather, C. H. Book in English, Greek, Ancient (to 1453). *Library of history: Bibliotheca historia*. Cambridge, MA: Harvard University Press.

Evanari, M., Leslie, S., & Naphtali, T. (1971). Man in the Negev. In M. Evanari, L. Shanan, & N. Tadmor (Eds.), *The Negev: The challenge of a desert* (pp. 11–29). Cambridge, MA: Harvard University Press.

Glueck, N. (1937). A newly discovered Nabataean temple of Atargatis and Hadad at Khirbet Et-Tannur, Transjordania. *American Journal of Archaeology, 41*(3), 361–376.

Graf, D. F., & Sidebotham, S. E. (2003). Nabataean trade. In G. Markoe (Ed.), *Petra rediscovered lost city of the Nabataeans* (pp. 65–74). Cincinnati: The Cincinnati Art Museum.

Kauth, R. J., & Thomas, G. S. (1976). The tassled cap: A graphic description of the spectral-temporal development of agricultural crops as seen by landsat. In *The laboratory for applications of remote sensing, Proceedings of the symposium on machine processing of remotely sensed data* (pp. 41–51). West Lafeyette: Purdue University Press.

Metz, H. C. (1989). *Jordan: A country study*. Washington, DC: GPO for the Library of Congress.

Negev, A. (1986). *Nabataean archaeology today* (Hagop Kevorkian series on near eastern art and civilization). New York: New York University Press.

Starky, J. (1955). The Nabataeans: A historical sketch. *The Biblical Archaeologist, 18*(4), 84–106.

Chapter 8
Beyond the Bend: Remotely Sensed Data and Archaeological Site Prospection in the Boyne Valley, Ireland

William Megarry and Stephen Davis

Abstract The UNESCO World Heritage Site of the Archaeological Ensemble of the Bend in the Boyne ('Brú na Bóinne') is one of the most important archaeological landscapes in Europe. This paper reviews results from an ongoing project which is using high-resolution elevation (LiDAR) data and Worldview2 multispectral imagery to identify and catalogue new sites in and around the World Heritage Site. It stresses the importance of using a range of methodologies when remotely exploring landscapes, and illustrates how even within one of the most studied landscapes in Europe remote sensing technologies can identify new and exciting information.

Keywords World heritage site • Landscape • Prospection • LiDAR • Multispectral

8.1 Introduction

The multi-period archaeological landscape of Brú na Bóinne ("The Bend in the Boyne") is widely regarded as one of the most important prehistoric archaeological landscapes in Europe (see Bradley 1998; Cooney 2000; Eogan 2012; Lewis-Williams and Pearce 2005 or O'Kelly 1998, 7). Situated in County Meath in the Republic of Ireland (Fig. 8.1), it contains Europe's largest and most important concentration of megalithic art (O'Sullivan 2006; Twohig 1981, 2000). The significance of the region was acknowledged in 1993 when the Archaeological Ensemble of the Bend in the Boyne was inscribed in the UNESCO World Heritage List. At the time of inscription, the UNESCO Committee stressed the importance of continual monitoring of future developments in and around the World Heritage Site (WHS). This proviso reflected the sensitive nature of the landscape which remains an intensely farmed agricultural region within commuting distance of major urban centres.

Results from ongoing research using remotely sensed data and imagery to monitor existing and inventory new archaeological sites within and around the boundary of the WHS are presented in this paper, which is written in four sections. The Bend

D.C. Comer and M.J. Harrower, *Mapping Archaeological Landscapes from Space*, SpringerBriefs in Archaeology, DOI 10.1007/978-1-4614-6074-9_8, © Springer Science+Business Media New York 2013

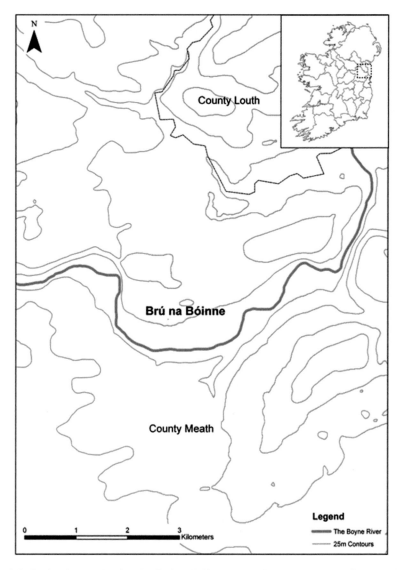

Fig. 8.1 Regional map showing the Brú na Bóinne or Bend in the Boyne in County Meath, Ireland

in the Boyne is introduced in the first section, which also provides a brief summary of major archaeological fieldwork in the WHS. Results from an intensive desktop LiDAR survey are presented in the second section, while the third section reviews how multispectral Worldview2 data was utilised to generate vegetation and moisture indices. These layers can help record subsurface archaeology. The final section explores potential future avenues for research in the region.

8.2   Regional Introduction

The situation of the archaeological complex on elevated ground surrounded on three sides by water gives the landscape an emotive and almost mythical feel experienced by the many thousands of visitors to the region each year. The ensemble is centred on three large tumuli (Newgrange, Knowth and Dowth) and contains at least 40 other smaller funerary and associated sites, spread across elevated carboniferous shale ridges running parallel to the modern river (Fig. 8.2). Both Newgrange and Knowth have been excavated in recent decades, leading to extensive and ongoing publication (Eogan 2012, 1984; Eogan and Roche 1997; Eogan and Doyle 2010; O'Kelly 1978; Stout 2002; Stout and Stout 2008). The three great tumuli are particularly noted for their solar alignments. At the winter solstice, the rising sun shines down through a specially-constructed roofbox, illuminating the chamber at Newgrange. In this cruciform-shaped chamber, three side recesses held burials in large stone basins. Elaborate designs were carved into the orthostats that form the passageway. Designs from these orthostats have become synonymous with Irish identity including the iconic *triskele* (triple spiral) from the northern recess at Newgrange.

Radiocarbon dating suggests that while initial settlers to the region arrived around 3900 B.C. during the Early Neolithic, the larger structures were constructed later, between 3,350 and 2,900 B.C. (Cooney 2000). They were situated at the heart of a

Fig. 8.2 Panchromatic Worldview2 image showing the core area of the WHS including the tumuli at Knowth, Newgrange and Dowth

long-lived and complex funerary and ritual landscape – designed to be both visited and seen from afar. The construction of these tombs marked the beginning of a process of monumentalising the landscape that was to continue for 5,000 years. Large circular earthen enclosures (some more than 150 m in diameter), standing stones and smaller passage tombs indicate continued constructions and occupation of the landscape throughout the prehistoric and into the historic period. The landscape was also the location of the Battle of the Boyne when, in 1690, Protestant Williamite forces defeated the Catholic King James. It is a landscape of both ancient and contemporary importance which retains significance to people across the island of Ireland.

The earliest historical texts from the region make both mythical and historical references to the complex. Brú na Bóinne was believed to be the primary resting place for members of the mythical *Tuatha Dé Danann* (people of the Goddess Danú) who ruled Ireland before the coming of the Gaels (O'Kelly 1978, 70). Other historical references to the burial of Cormac MacAirt, the third-century King of Tara, at the complex reflect a deep ancient appreciation of the significance and importance of the landscape. More modern interest in the landscape dates back more than 300 years when the then keeper of the Ashmolean museum Edward Lhywd visited the giant tumulus of Newgrange following the discovery of a passage by the landowner and Dublin merchant Charles Campbell. Early maps from the region reflect the cartographer's interest in antiquaries. The 6-in. Ordnance Survey maps of the region were drawn in the 1830s and record moats, rings, standing stones and forts in abundance spread across the contemporary agricultural landscape. While such surveys often failed to apply a correct nomenclature, they served a vital purpose in recording the landscape at a time before large scale development.

Many sites recorded by historical mapping have since disappeared, destroyed for building material or by intensive agriculture. The majority of excavation activity in the region took place in the late twentieth century, focusing on the large tumuli at Knowth and Newgrange. During these larger campaigns, excavators were keen to contextualise the tombs within their greater archaeological landscapes, and smaller sites too were recorded and mapped (see for example O'Kelly 1978; Eogan 1984). More recent field survey has further contextualised these sites in the Neolithic, recording concentrations of lithic scatters across the landscape (Brady 2005). In 2009 a detailed research framework was published which outlined key research questions, a research strategy and recommendations for the future management of the landscape (Smyth 2009). This report built upon earlier documents which reflected the chronologically fragmentary nature of the archaeological record in the Bend in the Boyne (O'Neill 1989; Duchas 2002; Heritage Council 2002).

8.3 LiDAR Survey

In 2008, a 3-year Geographical Information Systems (GIS) project was funded by the Irish Heritage Council. This project aimed to produce an integrated, comprehensive landscape archaeological model of the evolution of the Boyne Catchment, and

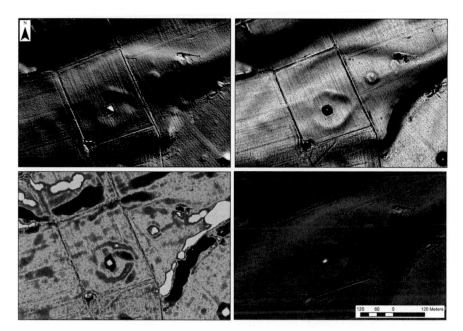

Fig. 8.3 Composite image showing LiDAR visualisation techniques. *Top left*: Hillshade model (Azimuth 315°, Inclination 45°). *Top right*: SkyView factor (Kokalj at al. 2011). *Bottom left*: Localised relief model (Hesse 2010). *Bottom right*: Multi-directional hillshade model (16 directions)

thus develop an environmentally-contextualised understanding of a key Irish archaeological landscape.

The project was undertaken in three phases. Phase 1 involved compiling a large amount of digital data including historical maps, aerial imagery and high resolution 1 m LiDAR while Phase 2 was interested in mapping the palaeoenvironmental evolution of the river and the surrounding landscapes (Lewis et al. 2009). Phase 3 began in 2010 and was focused explicitly on expanding our understanding of the archaeological landscape through site prospection using the LiDAR data. LiDAR is ideally suited to the Irish archaeological landscape as it captures the often subtle palimpsest of earthworks and scrapings which record this history of a landscape. Many sites in the Boyne had remained invisible to traditional fieldwalking or aerial photographic techniques because of their low relief, and analytical techniques were employed to exaggerate surface elevation values and emphasise anthropogenic activity. These included various hillshading models including multi-directional shading, localised relief models (LRM) which use focal statistics to emphasise local variation by removing background redundancy (Hesse 2010) and techniques which explored the amount of solar illumination across a surface like the SkyView Factor (Zakšek et al. 2011; for a good overall review of LiDAR visualization techniques see Challis et al. 2011) (Fig. 8.3).

The survey was primarily desktop-based but also involved periods of fieldwork and geophysical survey. More than 100 potential new sites were identified within a

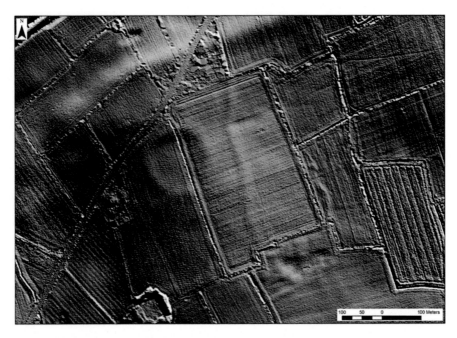

Fig. 8.4 Hillshade model showing previously unknown enclosure and earthworks at Carranstown south of the WHS (Azimuth 315°, Inclination 45°)

17×7 km survey grid. These included a number of new passage tombs in the core region of the WHS and several large enclosures (Davis et al. 2010) (Fig. 8.4). These results were even more impressive given the history of intense research in the region. Many new sites were located in close proximity to major monuments in areas which had been repeatedly investigated and explored. The LiDAR surface also enabled research into the landscape itself rather than individual sites. A complex of raised routeways was identified. Some of these were substantial double banked structures with clear associations with archaeological sites.

As previously stated, LiDAR data is ideal for detecting earthworks in the Irish landscape where monumental sites are still clearly visible beneath modern field systems. The technique proved far less effective identifying linear or angular remains like walls or buildings. This is partially because of the relatively coarse 1 m resolution of the dataset, but also reflects the nature of the archaeology. Earthworks are traditionally larger and have often been partially ploughed out, extending their topographical signature considerably. Archaeological landscapes are deeply complex phenomena and LiDAR only records a part of this complexity. LiDAR survey proved especially powerful when used alongside other spatial data like aerial photography and historical mapping. Results from the survey were also used to locate sites suitable for geophysical survey. Many of these sites had little or slight surface expression in the LiDAR but became clearly evident following ground-based geophysical survey. In some cases, new features were evident through geophysics alone, further emphasising the importance of a multifaceted remote sensing approach (Davis et al. 2012). Geophysical survey is an expensive and time-consuming activity unsuitable for large-scale landscape projects. A cost-effective technique was therefore sought to explore sub-surface activity invisible to the LiDAR data and traditional field survey.

8.4 Multispectral Imagery Survey

Aerial photography, both vertical and oblique, has been an invaluable tool for archaeologists seeking to understand the landscape from a distance (see for example Wilson 1982). While oblique images use lighting to detect topographical variation in a method similar to LiDAR, vertical images are often reliant on subtle crop marks representing variations in vegetation growth caused by subsurface features. In some cases, soil markings can also indicate subsurface features, especially in ephemeral conditions following rain, snow or frost. In some wet environments, archaeological features can affect the permeability of the soil resulting in drainage patterns, which in turn reflect underlying structures (Wilson 1982, 39–70). In such cases, it is often possible to detect subsurface architecture or ditches.

The problem with such vertical imagery is that is heavily dependent on environmental and meteorological factors and therefore difficult to predict and plan for data acquisition. Three phases of vertical aerial photography were available to the project (1995, 2000 and 2005) but these images lacked the detail necessary to properly explore the landscape. A single swathe of Multispectral Worldview-2 (MS WV2) imagery was obtained from DigitalGlobe. This data included eight spectral bands at 1.8 m resolution and a single panchromatic band at 0.5 m resolution. The higher resolution panchromatic band can be used to pan-sharpen the coarser colour bands, producing an eight band image of the entire landscape at 0.5 m resolution (Fig. 8.5). While traditional remotely sensed imagery like Landsat has favoured larger regional studies identifying extensive remains at larger scales (Comfort 1997; Comer in this volume), the high-resolution of the MS WV2 imagery allows for more detailed landscape studies at smaller scales. The MS WV2 imagery captures a snapshot of the landscape in multispectral detail and can be used to generate indices capable of displaying and quantifying variation in vegetation or moisture content within the soil. These indices are constant at that moment and can therefore be used to explore the spectral signature across the entire landscape. MS WV2 data is also atmospherically corrected minimising post-purchase processing. The impact of subsurface archaeology on crop growth was assessed through the generation of vegetation indices.

Vegetation indices measure plant health in satellite imagery. While directly relevant in ecology and related fields, their ability to reflect plant growth variation can be used to detect subsurface features. While normally only visible in ephemeral conditions, such indices can pick up crop marks in sharp detail. Two indices were generated from the 8-Band images: a Normalised Difference Vegetation Index (NDVI) generated using the Worldview2 red edge band, and a wetness index generated using a tasselled-cap transformation (Fig. 8.6). The NDVI returns a value between −1 (low)

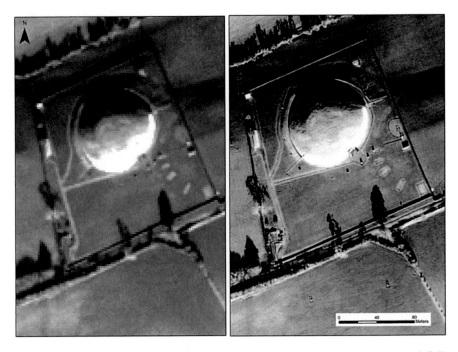

Fig. 8.5 RGB Worldview2 image showing the passage tomb at Newgrange: unsharpened (*left*) and pan-sharpened using the panchromatic band (*right*)

and 1 (high) reflecting the chlorophyll content of plants, while the wetness index explores the amount of moisture in the soil. A trial area was selected east of the Hill of Slane to the west of WHS boundary. This area had been previously noted as potentially significant from an historical archaeological perspective, given the presence of a suggested quarrying complex clearly visible in the LiDAR imagery. The features were also angular and difficult to isolate on the LiDAR. It was hoped that the aforementioned indices would detect further features reflected in the plant growth. Results were positive identifying linear features including a ramp, boundaries and probable buildings. Aside from vegetation indices, principle component analysis (PCA) was used to reduce redundancy across the bands (Parcak 2009). This produced some interesting results in the area around the passage tomb at Knowth.

8.5 Project Review and Avenues for Future Research

While the technological and processual steps in prospection using remotely sensed data and imagery are deeply quantitative, the technique remains highly qualitative in its interpretation. Both the LiDAR and MS WV2 surveys involved a two-step

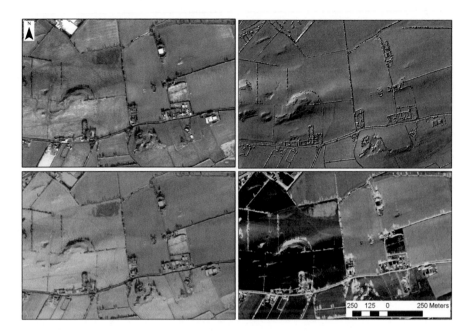

Fig. 8.6 Composite image showing the area east of Slane. *Top left*: Pan-sharpened Worldview2 RGB image. *Top right*: Lidar hillshade model (Azimuth 315°, Inclination 45°). *Bottom left*: Pan-sharpened NDVI index generated using Worldview2 *red* and *red* edge bands. *Bottom right*: Tasseled-cap transformation showing wetness (Coefficients from Yarborogh and Ravi *forthcoming*)

process where data was subject to a variety of modelling techniques before being visually examined. The latter part of the process is very time-consuming and subjective. By developing a robust statistical sampling model, it would be possible to automate the surveying of larger areas (see Menze and Ur in this volume). The project would also benefit from using MS WV2 imagery from different times of the year. Environmental factors can greatly influence the manifestation of subsurface archaeology in vegetation and crop marks. Features may also become visible by performing change detection between images from different times of the year.

8.6 Conclusion

We have shown how, even in an intensely studied archaeological landscape such as Brú na Bóinne, modern remote sensing techniques can bring a new layer of complexity to our understanding of the archaeological landscape. While at this stage, LiDAR is becoming widely established as a methodology in archaeological prospection (Opitz and Cowley 2012), to date multispectral satellite imagery has been largely the preserve of forestry and agriculture. However, it clearly offers a cost-effective (in comparison to LiDAR or airborne Hyperspectral imagery) and potentially valuable source of archaeological information which warrants further study. While it is a beginning to explore the range of indices which already exist to deal with these data, far greater study is required to understand which methodologies will prove effective in any given circumstance. The challenge then is to take such remote sensing beyond merely archaeological prospection and towards providing a better informed understanding of archaeological landscapes, both from a theoretical and a management perspective.

Acknowledgments The authors would like to thank and acknowledge the Heritage Council of Ireland for its generous support under the INSTAR scheme. During this project, valuable insight was also provided by Dr. Conor Brady (DKIT), Dr. Helen Lewis (UCD) and Dr. Jonathan Turner (UCD). Kevin Barton undertook the geophysical survey. The authors would also like to acknowledge Digital Globe who provided the Worldview2 data at part of their 8 Band Challenge.

References

Bradley, R. (1998). *The significance of monuments*. London: Taylor and Francis.
Brady, C. (2005). *Earlier prehistoric settlement in the Boyne Valley: A systematic ploughzone survey*. Unpublished Ph. D. thesis, University College Dublin.
Chadburn, A., & Pomeroy-Kellinger, M. (Eds.). (2001). *Archaeological research agenda for the Avebury world heritage site*. Salisbury: Wessex Archaeology for English Heritage.
Challis, K., Forlin, P. and Kincey, M. 2011. A Generic Toolkit for the Visualization of Archaeological Features on Airborne LiDAR Elevation Data. *Archaeological Prospection* 18, 279–289.
Comfort, A. (1997). Satellite remote sensing and archaeological survey on the Euphrates. *Archaeological Computing Newsletter, 48*, 1–8.
Cooney, G. (2000). *Landscapes of Neolithic Ireland*. London: Routledge, Apologies.
Davis, S., Brady, C., Megarry, W., & Barton, K. (2012). LiDAR survey and the Brú na Bóinne world heritage site. In R. Opitz & D. Cowley (Eds.), *Interpreting archaeological topography: 3D data, visualisation and observation*. Oxford: Oxbow.
Davis, S., Megarry, W., Brady, C., & Lewis, H et al. (2010). *Boyne valley landscapes project: Phase III Final Report 2010*. Dublin: Heritage Council.
Dúchas The Heritage Service and Department of the Environment and Local Government. 2002. *Brú na Bóinne World Heritage Site Management Plan*. Dublin: The Stationary Office.
Eogan, G. (1984). *Excavations at knowth I: Smaller passage tombs, Neolithic occupation and beaker activity*. Dublin: Royal Irish Academy.
Eogan, G. (2012). *Excavations at knowth volume 5: The archaeology of knowth in the first and second millenia A.D.*. Dublin: Royal Irish Academy.
Eogan, G., & Doyle, P. (2010). *Guide to the passage tombs at Brú na Bóinne*. Dublin: Wordwell Press.
Eogan, G., & Roche, H. (1997). *Excavations at knowth 2: Settlement and ritual sites of the fourth and third millennia B.C.*. Dublin: Royal Irish Academy.
Heritage Council (2002). *Brú na Bóinne Management Plan*. http://www.meath.ie/LocalAuthorities/Publications/HeritagePublications/File,35793,en.pdf. Accessed 31 March 2011.
Hesse, R. 2010. LiDAR-derived Local Relief Models – a new tool for archaeological prospection. *Archaeological Prospection* 17, 67–72.
Kokalj, Ž., Zakšek, K. & Oštir, K. 2011. Application of sky-view factor for the visualization of historic landscape features in lidar-derived relief models. *Antiquity* 85, 263–273.

Lewis, H., Gallagher, C., Davis, S., Turner, J., Foster, G., et al. (2009). *An integrated, comprehensive GIS model of landscape evolution and land use history in the River Boyne Valley: Phase 2.* Dublin: Heritage Council.

Lewis-Williams, D. J., & Pearce, D. G. (2005). *Inside the Neolithic mind: Consciousness, cosmos and the realm of the gods.* London: Thames and Hudson.

O'Neill, A. (1989). *National Archaeological Park: Boyne Valley.* Unpublished Consultancy Report.

O'Sullivan, M. (2006). The Boyne and beyond: A review of megalithic art in Ireland. In R. Joussaume, L. Laporte, & C. Scarre (Eds.), *Origin and development of the megalithic monuments of Western Europe* (pp. 651–688). Bougon, France: Le musée des Tumulus de Bougon.

O'Kelly, C. (1978). *Illustrated guide to newgrange and the other Boyne monuments.* Wexford: John English.

O'Kelly, M. J. (1998). *Newgrange: Archaeology, art and legend.* London: Thames and Hudson.

Opitz, R., & Cowley, D. (Eds.). (2012). *Interpreting archaeological topography: 3D data, visualisation and observation.* Oxford: Oxbow.

Parcak, S. H. (2009). *Satellite remote sensing in archaeology.* London/New York: Routledge.

Shee Twohig, E. (1981). *The megalithic art of Western Europe.* Oxford: Clarendon.

Shee Twohig, E. (2000). Frameworks for the megalithic art of the Boyne Valley. In G. Johnson, M. McCarthy, J. Sheehan, & E. Shee Twohig (Eds.), *New agendas in Irish prehistory: Papers in commemoration of Liz Anderson* (pp. 89–105). Bray: Wordwell.

Smyth, J. (2009). *Brú na Bóinne world heritage site: Research framework.* Dublin: The Heritage Council.

Stout, G. (2002). *Newgrange and the bend in the Boyne.* Cork: Cork University Press.

Stout, G., & Stout, M. (2008). *Newgrange.* Cork: Cork University Press.

Wilson, D. R. (1982). *Air photo interpretation for archaeologists.* London: Batsford.

Chapter 9
Archaeological Remote Sensing in Jordan's Faynan Copper Mining District with Hyperspectral Imagery

Stephen H. Savage, Thomas E. Levy, and Ian W. Jones

Abstract Hyperspectral (multiple, narrow band) satellite imaging provides a useful discovery and analytical tool for archaeologists. The Hyperion instrument, flying on the Earth Observer 1 (EO-1) satellite, was launched from Vandenberg Air Force Base on November 21, 2000. Hyperion provides 242 (196 calibrated) narrow bands in the visible (VIS) to shortwave infrared range (SWIR), enabling detailed archaeological and geological analyses. We analyzed a Hyperion image swath targeted on Khirbat en-Nahas (KEN), an ancient copper smelting site along the Wadi al-Ghuwayb (WAG) in Jordan's Faynan district, where extensive ore processing occurred from the third millennium B.C.E. to industrial scale production over several centuries in the early first millennium B.C.E. (Iron Age) and continued until Medieval Islamic times. We use a combination of Principal Components Analysis (PCA), similarity matrices, and Spectral Mixture Analysis (SMA) to locate additional ore processing sites and discern depositional differences that may help illuminate issues related to the organization of production at KEN. Extensive field surveys in the research area provide a unique opportunity to "ground truth" the results of the hyperspectral research. Our results show considerable promise for future work with Hyperion

Keywords Hyperion • Iron age copper mining • Jordan • Spectral mixture analysis • Principle components analysis

9.1 Introduction

Satellite remote sensing in archaeological research has most often utilized imagery from sources such as Landsat, ASTER, SPOT, and CORONA. A wide-range of work shows the utility of these well-known resources; some of them have been applied to ancient mining and smelting sites (Deroin et al. 2011; Pryce and Abrams 2010). But the use of hyperspectral imagery is less well known, with some exceptions (Traviglia 2006; Parcak 2009:101–102; Kwong et al. 2009).

D.C. Comer and M.J. Harrower, *Mapping Archaeological Landscapes from Space*, SpringerBriefs in Archaeology, DOI 10.1007/978-1-4614-6074-9_9, © Springer Science+Business Media New York 2013

Fig. 9.1 The western part of
the Faynan District covered
by the Hyperion image. The
image area is shown as the
red rectangle in the inset

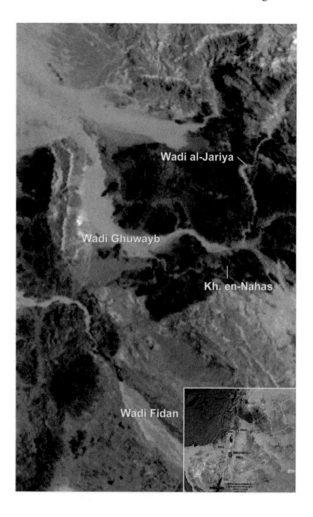

The use of space-based hyperspectral platforms, such as NASA's Hyperion
instrument, originally developed especially for geological and mining applications,
has only recently begun to be explored (Alexakis et al. 2009). This chapter summarizes the results of a study that is more comprehensively described in Savage
et al. (2012). We present an analysis of the copper mining region in the Faynan
District of southwestern Jordan (Fig. 9.1), an area that has been intensively studied
by archaeologists since the early 1980s (e.g., Barker et al. 2007; Hauptmann 2007;
Levy et al. 2008). Principal Components Analysis (PCA), construction of similarity
matrices, and spectral mixture analysis of geologically-based endmembers were
used to examine a Hyperion image that includes the Iron Age copper smelting site
of Khirbat en-Nahas (KEN; Arabic=ruins of copper) – the largest Iron Age copper
production site in the southern Levant (Figs. 9.2 and 9.3, Levy et al. 2004, 2005,
2012). In 2005, 2006, and 2009 large-scale excavations were carried out at KEN

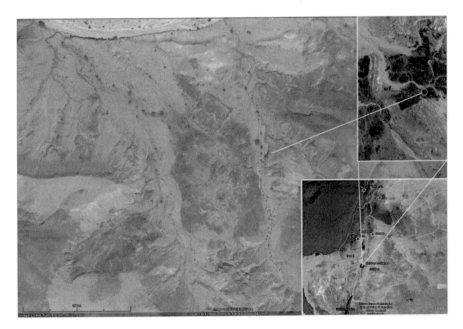

Fig. 9.2 Google earth image of Khirbat en-Nahas. (The *dark gray* areas are slag mounds; the square structure is an Iron Age fortress. The *bottom right* inset shows the location of the Hyperion study area; the *top right* inset shows the hyperion image with the location of Khirbat en-Nahas *circled in yellow*)

Fig. 9.3 Overview of the Iron Age (ca. 1,200–900 B.C.E.) copper production site of Khirbat en-Nahas, southern Jordan. *Extensive black* 'slag mounds' indicate ancient smelting of copper ore mined locally (Photo: T.E.L, UCSD Levantine Archaeology Laboratory)

that focused on systematically excavating one of the deeply stratified "slag mounds" at the site and other areas (Levy et al. 2008).

9.2   Geology of the Faynan District

The Faynan District is located in southwestern Jordan, between north latitudes 30–31 and east longitude 35–36. Faynan is part of the Wadi Araba rift system, which is a northern extension of the great East African Rift Valley. Rabb'a (1994) presents a detailed description of the geology of the Faynan District, which is only touched upon here; Hauptmann (2007) provides a more general description. According to Rabb'a, "Copper mineralization is present in the volcanic acidic rocks of the Ahaymir Suite as crusts infilling joints but the main Cu-ore targets are in the sedimentary rocks unconformably overlying the basement Araba complex units… The second most common occurrence is present in the Burj Dolomite-Shale (BDS) Formation which comprises three members from base to top, the Tayan Siltstone, Numayri Dolomite and Hanneh Siltstone. The most important member for copper mineralization is the Numayri Dolomite, which is about 40 m thick" (Rabb'a 1994: 47). BDS occurs in large deposits north of the Wadi al-Ghuwayb, on both sides of its tributary, the Wadi al-Jariya, and in isolated pockets south of the Wadi al-Ghuwayb (Fig. 9.4)

9.3   The Earth Observer 1 (EO-1) Satellite and the Hyperion Instrument

The NASA EO-1 satellite was launched on November 21, 2000 on a 1-year technology validation/demonstration mission. The EO-1 spacecraft carries the Hyperion instrument, the first relatively high (30 m) spatial resolution imaging spectrometer to orbit the earth. Hyperion has 242 (196 are calibrated) spectral bands in the range 0.4–2.5 µm, in the Visible (VIS) and Short Wave Infrared (SWIR) ranges. Of the 196 calibrated bands, 40 others exhibit considerable streaking, caused by calibration differences in the detector array, which vary according to ground and atmospheric conditions (Goodenough et al. 2003: 1322–1323). This left 156 viable bands for the present analysis. We used the Davinci software application to process the Hyperion data swath.[1]

---

[1] Davinci is a free, open source application developed at the Mars Space Flight Facility, School of Earth and Space Exploration, Arizona State University. The application runs under Windows and UNIX, and can be downloaded from http://davinci.mars.asu.edu/index.php. In addition, we used Global Mapper to create image overlays; in each the end product of various steps in Davinci was added as one or more layers to three visible light bands of the Hyperion image.

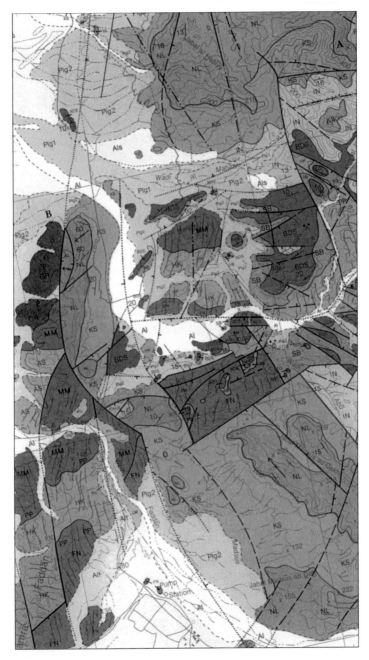

Fig. 9.4 Geology of the Faynan Region (From Rabb'a 1994). *HK* Hunayk Monzogranite, *FN* Feldspar Porphyry, *BDS* Burj Dolomite-Shale, *IN* Umm Ishrin Sandstone, *KS* Kurnub Sandstone

9.4   Principal Components Analysis and the Search for Ore Processing Sites

The slag mounds at Khirbat en-Nahas are easily seen in the visible light bands of the Hyperion image; the mounds at the site are large, spread over a substantial area, and stand out prominently from the surrounding surfaces. Other ore processing sites in the region are not as large, and hence, not as easy to see in the visible satellite image. But there may be other band combinations beyond the visible spectrum that can make such places stand out. Principal Components Analysis (PCA) (Lillesand et al. 2008: 527–533) is a dimension reduction technique that preserves the variability in the original dataset. It has been used successfully in several archaeological remote sensing applications (see Parcak 2009: 97–99).

Figure 9.5 illustrates a portion of the study area (enlarged for later discussion) showing the first component as blue, the second as green, and the fourth as red. Together, these three components encompass 98.75% of the variance in the original 156 bands. Khirbat en-Nahas clearly stands out as a group of orange pixels of different intensities against a background of different shades of green; it is the area shown in the white rectangle near the bottom center of the figure. We found that the PCA procedure successfully identified areas where mineshafts and ore processing sites had been previously identified through archaeological survey (the other areas in Fig. 9.5 delineated by white rectangles). In addition, the PCA procedure has illuminated differences in the spectral signatures among the various pixels at KEN that represent slag mounds. These observations can be more fully addressed by constructing similarity matrices and spectral mixture analysis, to which we now turn.

9.5   A Similarity Matrix for Khirbat en-Nahas Slag Mounds

A closer examination of the pixels that represent slag mounds at KEN shows different color values (shades of orange) indicated in areas of the site represented by different slag mounds, and at the various hotspots discussed above. Other pixels in the Hyperion image cube that have a very similar spectral signature to specific pixels at the site would indicate areas where ores were extracted and processing activities are more likely to occur. We can count the number of bands whose value is the same for any slag mound location at KEN using Davinci. This results in an image with values ranging from 0 to 156, where the higher numbers represent pixels that are more similar to the source pixel. Figure 9.6 illustrates the similarity matrices for six loci at KEN that represent slag mounds. For each locus, we present only those places in the rest of the study area where at least two-thirds of the band values are identical. Most of the high similarity regions are located north of the site, across Wadi al-Ghuwayb and up the Wadi al-Jariya, though there are also a number of similar pixels in the region of the Khirbet en-Nahas mines southwest of the main site. The similarity matrix picks up the mapped concentrations of the main copper ore bearing

Fig. 9.5 Results of principal components analysis on the hyperion image cube, showing the site and areas with similar spectral signatures. Lower case letters refer to areas discussed in the text of Savage et al. (2012)

deposit in the region – the Burj Dolomite Shale unit, and in particular, the northern portion of the Wadi al-Jariya where more than 70 mines have been found (Levy et al. 2003; Knabb et al. in prep). However, the similarity matrix also highlights areas dominated by different kinds of granite, which indicates that the slag mounds

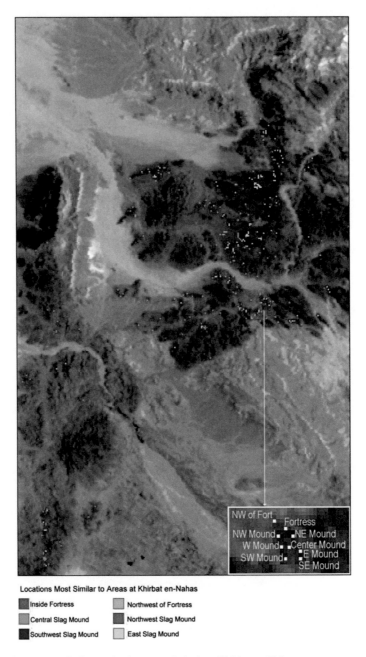

Fig. 9.6 Areas most similar to six slag mound pixels at Khirbat en-Nahas

contain mixed spectral signatures. We will use spectral mixture analysis below to unmix these areas.

9.6 Spectral Mixture Analysis

Rather than placing each pixel in an image into one of a set of discrete classes, as conventional image classification does, spectral mixture analysis (SMA) starts with a set of endmembers, representing a spectral library, and then deconvolves the image through linear unmixing to reveal the percentages of endmembers present in each pixel analyzed (see Ramsey and Christensen 1998). Spectral endmembers are defined as spectrally "pure" features (e.g. vegetation, soil, etc.) usually under idealized in situ or laboratory conditions. When in situ measurements are not possible, spectral endmembers can also be derived from "pure" features in the imagery (Piwowar et al. 1999). Identification of endmembers is critical to the SMA operation. In this case, we used the geology map published by Rabb'a (1994) to identify 15 endmembers, based on the major classifications provided on the map. Figure 9.7 shows the locations of the end-member pixels chosen as red squares with abbreviations from Rabb'a (1994).

Khirbat en-Nahas contains more than 21 slag mounds, which have different spectral signatures and probably contain the detritus from different sources of ore and metallurgical activities. We conducted an SMA on pixels that represent six of the main slag mounds at the site. Table 9.1 summarizes the results, and the deconvolution of the center slag mound is shown in Fig. 9.8. The SMA shows that the various slag mounds visible in the Hyperion imagery have different end member mixtures from different sources. This suggests they may have been loci of different activities at the site, and indicates that some areas may have been more intensively used for ore processing than others (for example, the east and southeast mounds, which have only BDS present). Whether the mounds represent smelting activity that occurred at different times, or smelting that was conducted at broadly similar times but by different groups, remains an open question.

Burj Dolomite Shale dominates all the slag mound pixels tested, with the exception of the center mound. SMA indicates that the center slag mound is a mixture of four endmembers: Burj Dolomite Shale, Fluviatile Gravels (Plg2), Hunayk Granodiorite (HK) and Salib Arkosic Sandstone (SB), which is the rock upon which Khirbat en-Nahas rests. BDS is the main constituent of the copper-bearing ores smelted at the site, and the components in the central slag mound suggest that BDS was obtained from regions with the other, non ore-bearing materials. Since the similarity matrix indicated that there were areas quite similar to the central slag mound located in HK rocks near the JAJ mines northwest of the site, the HK material found in the SMA analysis may have originated there.

Our study provides intriguing insights into the possible organization of copper production in the Iron Age Faynan. If a single social group controlled access to the entire region's ore resources and to the entire processing site at KEN, then we might expect

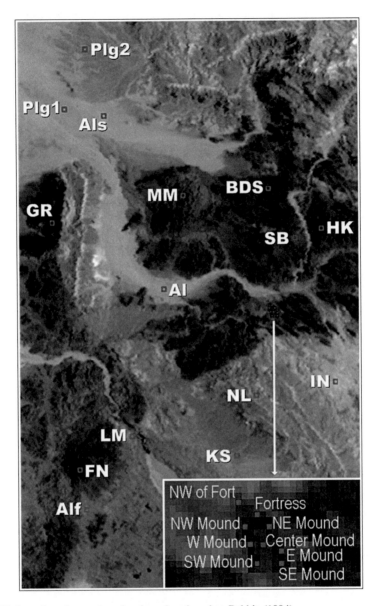

Fig. 9.7 Location of scene-based endmembers based on Rabb'a (1994)

to see little or no difference in the spectral signatures of the various slag mounds. Under these social conditions, we might expect that ore from any mine might be processed at any locus at KEN, which would create a more or less homogenous spectral signature across the entire site, with SMA endmembers from each of the different ore sources found in relatively equal amounts in each of the slag mounds examined. The same situation could be expected to occur if different social groups controlled different ore

Table 9.1 Results of spectral mixture analysis of Khirbat en-Nahas slag mounds

Mound	Endmembers and percentages present[a]
Northwest mound	BDS 90%, Plg2 10%
Northeast mound	BDS 95%, MM 5%, Plg2 <5%
East mound	BDS 100%
Southeast mound	BDS 100%
Southwest corner	BDS 75%, MM 15%, GR 5%, HK 5%
West mound	BDS 95%, Plg2 5%
Center mound	BDS 50%, Plg2 30%, HK 15% SB 10%
Fortress	BDS 90%, Plg2 10%, MM <5%
NW of Fortress	BDS 85%, AL 10%, Als 5%, MM <5%

[a]Rounded to nearest 5%; total may exceed 100%

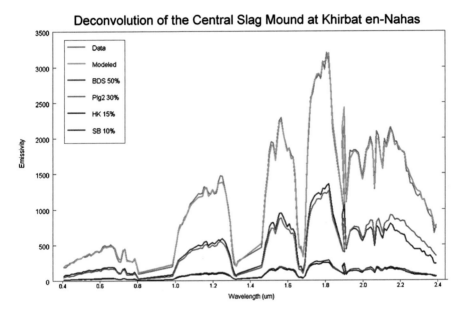

Fig. 9.8 Spectral mixture analysis deconvolution of the central slag mound at Khirbat en-Nahas

sources, but the processing was organized and run as a single enterprise – we would see ores from any or all of the different sources distributed more-or-less evenly across the entire production site. However, if both the ore sources and the processing were differentially controlled, then we might expect to see different ore sources being processed in different places at the site. That a single, larger entity (as represented by the fortress) may have had overall control of the copper production does not preclude the possibility that production was organized as a group of parallel operations, each drawing from its own resources, since the organization could easily be horizontal at the lower level, yet feed into a more vertically oriented state-run operation.

The results of our studies with similarity matrices and SMA indicate that the six pixels analyzed have different spectral signatures, corresponding to different mixtures of geological endmembers drawn form a variety of loci in the Faynan region. This suggests that a number of potentially independent ore processing operations, each with its own mines and smelting loci, operated at the site. However, because of the coarse resolution of the Hyperion imagery, we could not perform the analysis on each of the 21 major slag mounds present at KEN because we cannot distinguish them in the image. Higher resolution imagery will help resolve these questions, which can be further addressed through excavations of a variety of the slag mounds, comparing the slag to the ore signatures.

9.7   Summary and Evaluation

Hyperion imagery presents a new and useful resource to archaeologists. The preliminary studies reported here indicate that techniques such as PCA, similarity matrices, and SMA can provide much additional information that is of archaeological interest, opening up new avenues for potential research and investigation. Through these techniques we have shown where there are areas in the Hyperion image that may indicate additional ore extraction or smelting activities, and we have shown that the components of the slag mounds at Khirbet en-Nahas have different mixes of geological endmembers. That result suggests that different areas of the site were used to process ores from different sources, rather than all areas of the site being used to process ores from all of the different mines, which helps open up additional avenues of inquiry into the social relations of production in the Iron Age Faynan.

Acknowledgements We express our thanks to several people at NASA, including Stuart Frye, Jay Pearlman, and Stephen Ungar, for their help in setting up a Hyperion data acquisition request, and answering many questions about the nature of the Hyperion data product. This study could not have been done without them. Thanks are also due to Alicia Rutledge and Christopher Edwards, Research Assistants at the ASU Mars Space Flight Facility, for introducing us to the Davinci software, and especially to Chris, who helped develop atmospheric correction and similarity matrix routines. We owe a great deal to Dr. Phil Christensen, for two semesters of graduate level remote sensing lectures, for explaining the intricacies of the Spectral Mixture Analysis routines in Davinci, and for reading and commenting on an earlier draft of the paper. Thanks to Ramesh Rao, Director, Calit2 San Diego Division; Ziad al-Saad, former Director General, Department of Antiquities of Jordan; Mohammad Najjar, UCSD Levantine Archaeology Lab.

References

Alexakis, D., Sarris, A., Astaras, T., & Albanakis, K. (2009). Detection of neolithic settlements in Thessaly (Greece) through multispectral and hyperspectral satellite imagery. *Sensors, 9*, 1167–1187.

Barker, G., Gilbertson, D., & Mattingly, D. (Eds.). (2007). *Archaeology and desertification: The wadi faynan landscape survey, Southern Jordan*. Oxford: Council for British Research in the Levant and Oxbow Books.

Deroin, J., Téreygeol, F., & Heckes, J. (2011). Evaluation of very high to medium resolution multispectral satellite imagery for geoarchaeology in arid regions: Case study from Jabali, Yemen. *Journal of Archaeological Science, 38*(2011), 101–114.

Goodenough, D. G., Dyk, A., Niemann, K. O., Pearlman, J. S., Chen, H., Han, T., Murdoch, M., & West, C. (2003). Processing hyperion and ALI for forest classification. *IEEE Transactions on Geoscience and Remote Sensing, 41*(6), 1321–1331.

Hauptmann, A. (2007). *The archaeo-metallurgy of copper: Evidence from faynan, Jordan.* New York: Springer.

Knabb, K., Levy, T. E., Najjar, M., & Jones, I. W. N. (in prep). Patterns of Iron Age mining and Iron Age settlement in Jordan's Faynan District: The Wadi al-Jariya survey in context. In T. E. Levy, E. Ben-Yosef, & M. Najjar (Eds.), *New insights into the Iron Age archaeology of edom, Southern Jordan: Surveys, excavations and research from the Edom Lowlands Regional Archaeology Project(ELRAP) 2006–2008.* Boston: Annual of the American Schools of Oriental Research.

Kwong, J. D., Messinger, D. W., & Middleton, W. D. (2009). Hyperspectral clustering and unmixing for studying the ecology of state formation and complex societies. *Imaging Spectrometry* XIV, Article # 7457 0E.

Levy, T. E., Adams, R. B., Anderson, J. D., Najjar, M., Smith, N., Arbel, Y., Soderbaum, L., & Muniz, M. (2003). An Iron Age landscape in the edomite lowlands: Archaeological surveys along wadi al-ghuwayb and wadi al-jariya, jabal hamrat fidan, Jordan, 2002. *Annual of the Department of Antiquities Jordan, 47*, 247–277.

Levy, T. E., Adams, R. B., Najjar, M., Hauptmann, A., Anderson, J. A., Brandl, B., Robinson, M., & Higham, T. (2004). Reassessing the chronology of biblical Edom: New excavations and 14C dates from khirbat en-nahas (Jordan). *Antiquity, 78*, 863–876.

Levy, T. E., Ben-Yosef, E., & Najjar, M. (2012). New perspectives on Iron Age copper production and society in the Faynan Region, Jordan. In V. Kassianidou, & G. Papasavvas (Eds.), *Eastern Mediterranean metallurgy and metalwork in the 2nd millennium B.C.* Oxford: Oxbow Books.

Levy, T. E., Higham, T., Bronk Ramsey, C., Smith, N. G., Ben-Yosef, E., Robinson, M., Münger, S., Knabb, K., Schulze, J. P., Najjar, M., & Tauxe, L. (2008). High-precision radiocarbon dating and historical biblical archaeology in southern Jordan. *Proceedings of the National Academy of Sciences, 105*, 16460–16465.

Levy, T. E., Najjar, M., van der Plicht, J., Smith, N. G., Bruins, H. J., & Higham, T. (2005). Lowland Edom and the high and low chronologies: Edomite state formation, the bible and recent archaeological research in Southern Jordan. In T. E. Levy & T. Higham (Eds.), *The bible and radiocarbon dating: Archaeology, text and science* (pp. 129–163). London: Equinox Publishing.

Lillesand, T., Keifer, R., & Chipman, J. (2008). *Remote sensing and image interpretation* (6th ed.). Hoboken: Wiley.

Parcak, S. H. (2009). *Satellite remote sensing for archaeology.* New York: Routledge.

Piwowar, J. M., Peddle, D. R., & Davidson, D. P. (1999). Assessing annual forest ecological change in Western Canada using temporal mixture analysis of regional scale AVHRR imagery over a 14 year period. In *Proceedings, 4th international airborne remote sensing conference and exhibition/21st Canadian symposium on remote sensing*, Vol. II (pp. 91–97), Ottawa, 21–24 June 1999.

Pryce, T., & Abrams, M. (2010). Direct detection of Southeast Asian smelting sites by ASTER remote sensing imagery: Technical issues and future perspectives. *Journal of Archaeological Science, 37*, 3091–3098.

Rabb'a, I. (1994). Geological of the Al Qurayqira (Jabal Hamra Faddan Area), Map Sheet No. 305111. 1:50,000 Geological Mapping Series Geological Bulletin No 28. Amman: Geology Directorate, Department of Natural Resources.

Ramsey, M. S., & Christensen, P. R. (1998). Mineral abundance determination: Quantitative deconvolution of thermal emission spectra. *Journal of Geophysical Research, 103*, 577–596.

Savage, S. H., Levy, T. E., & Jones, I. W. (2012). Prospects and problems in the use of hyperspectral imagery for archaeological remote sensing: A case study from the faynan copper mining district, Jordan. *Journal of Archaeological Science, 39*, 407–420.

Traviglia, A. (2006). Archaeological usability of hyperspectral images: Successes and failures of image processing techniques. In S. Campana, & M. Forte (Eds.), *From space to place: Proceedings of the 2nd international conference on remote sensing in archaeology,* BAR International Series. 1568 (pp. 123–130). Oxford: Archaeopress, 4–7 Dec 2006.

Part III
SAR (Synthetic Aperture Radar)

Chapter 10
Synthetic Aperture Radar, Technology, Past and Future Applications to Archaeology

Bruce Chapman and Ronald G. Blom

Abstract Synthetic Aperture Radar (SAR) data are now widely available at resolutions better than 30 m for most locations on Earth. This imagery can be quite valuable for archaeological applications in diverse environments, as it is generally unaffected by cloud cover, making it useful in the tropics. In very arid environments, longer wavelength radar can penetrate thin sand cover.

Radar imagery is sensitive to evidence of human occupation, such as remains of walls. Multi-temporal radar coverage, and/or coverage at multiple wavelengths and polarizations are available for some areas, from which vegetation characteristics can be derived. Depending on observational parameters, Interferometric Synthetic Aperture Radar (InSAR) data can provide topographic data, or measure subtle topographic changes occurring in the time interval between observations.

SAR data has been found useful for a variety of archaeological applications, with future instruments and technology promising expanded usage. Applications have ranged from airborne and spaceborne studies of Angkor Wat, to the Egyptian Sahara, to focused studies at locations such as San Clemente Island. For archaeologists, successful analysis of the data requires an understanding of basic principles of SAR image interpretation, knowledge of data sources, and essential sensor characteristics. Fortunately, radar data is available for download through various websites, and there are many software tools available, some available at no charge, to assist with obtaining data, analysis and interpretation.

Keywords SAR • Synthetic aperture radar • Remote sensing • UAVSAR • AIRSAR • ALOS • SIR-A • SIR-B • Desdyni • TandemX • Subsidence • Interferogram • Polarimetry • Soil penetration • Digital elevation models • Archaeology • Radar rivers • San Clemente Island • Angkor Wat

D.C. Comer and M.J. Harrower, *Mapping Archaeological Landscapes from Space*,
SpringerBriefs in Archaeology, DOI 10.1007/978-1-4614-6074-9_10,
© Springer Science+Business Media New York 2013

10.1   Introduction

Synthetic Aperture Radar (SAR) data are now widely available at resolutions better than 30 m for most locations on Earth. This imagery can be quite valuable for archaeological applications in diverse environments, as it is generally unaffected by cloud cover making it useful in the tropics; and in very arid environments, longer wavelength radar can penetrate thin sand cover.

Radar imagery is sensitive to evidence of human occupation such as remains of walls. Multi-temporal radar coverage, and/or coverage at multiple wavelengths and polarizations are available for some areas, from which vegetation characteristics can be derived. Depending on observational parameters, Interferometric Synthetic Aperture Radar (InSAR) data can provide topographic data, or measure subtle topographic changes occurring in the time interval between observations.

SAR data has been found useful for a variety of archaeological applications, with future instruments and technology promising expanded usage scenarios (see, for example, Golden and Davenport, this volume; Hixson, this volume; and Comer and Blom, this volume). Applications have ranged from airborne and spaceborne studies of Angkor Wat (Moore et al. 2007), to the Egyptian Sahara (Blom et al. 1984), to focused studies at locations such as San Clemente Island (Comer and Blom 2007). For archaeologists, successful analysis of the data requires an understanding of basic principles of SAR image interpretation, knowledge of data sources, and essential sensor characteristics. Fortunately, radar data is available for download through various websites, and there are many software tools available, some available at no charge, to assist with obtaining data, analysis and interpretation.

10.2   Examples of the Use of Synthetic Aperture Radar in Archaeology

10.2.1   Observing Sub-Surface Features in Arid Environments

In 1982, the second flight of the NASA Space Shuttle carried SIR-A (Shuttle Imaging Radar-A), an L-band imaging radar. Radar images from this instrument of the Sahara, now one of the driest places on Earth, showed a remarkable terrain carved by running water (see Fig. 10.1). Subsequent analysis and field work (McCauley et al. 1982) showed the current aeolian sand cover was largely transparent to the radar, revealing now dry watercourses from a wetter time. Archaeologically, this proved very significant in that concentrations of stone tools were associated with the "Radar Rivers."

Note that the resolution of the SIR-A images (~50 m) was inadequate to directly detect evidence of occupation, but it is more than adequate to indicate where people would have concentrated. The necessary conditions for sub-surface radar imaging

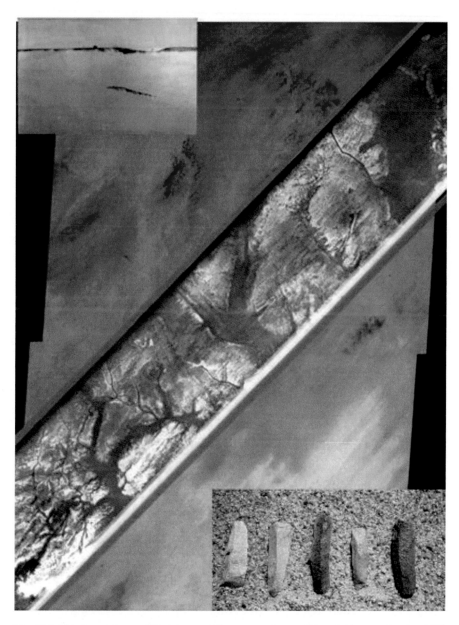

Fig. 10.1 This figure shows a SIR-A image of the western desert of Egypt laid upon a Landsat MSS image. The VNIR Landsat image shows the present day surface sand cover, while the radar image shows a fluvial landscape from wetter climatic episodes in the Sahara. *Upper left* inset shows from the ground the appearance of a "radar river". Note jeep on riverbank for scale. *Lower right* inset shows typical artifacts collected. SIR-A image swath width is 50 km (Elachi et al. 1984)

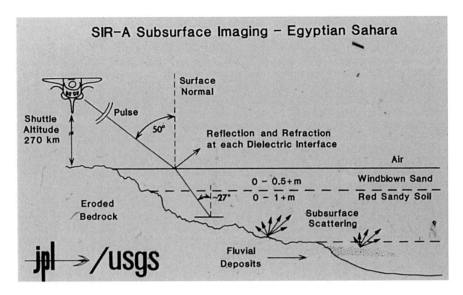

Fig. 10.2 Schematic of subsurface imaging in hyper arid terrain. Extremely dry sand and soil is relatively transparent, allowing radar energy to penetrate to more reflective substrate (bedrock and coarser fluvial deposits), which forms the SIR-A image. (Elachi et al. 1984)

are quite restrictive (Blom et al. 1984; Elachi et al. 1984; Farr et al. 1986) generally the cover material must be less than 2 or 3 m thick, homogeneous and fine grained relative to the radar wavelength (windblown sand is ideal), and extremely dry. Beneath this cover material must be a radar reflective landscape, or the radar image will not reveal subsurface features (see Fig. 10.2). For a more recent example, see Paillou et al. (2009).

10.2.2 Application of Multiple Polarizations and Digital Elevation Models in a Vegetated Environment

Angkor Wat was a major religious and urban center from the ninth through the fifteenth centuries in what is Cambodia today. Multi-polarization radar images from SIR-C (a radar mission flown in 1994 on NASA's Space Shuttle, Evans et al. 1997) and, later, DEMs from a NASA airborne SAR called AIRSAR (see AIRSAR 2012) made significant contributions to understanding Angkor's history (Moore et al. 2007). Continuing research, particularly by the Greater Angkor Project (see Greater Angkor Project 2012), enabled discovery of new structures and significantly enhanced regional understanding. The extensive urban complex of Greater Angkor developed over time with an increasingly elaborate water management system. By the fifteenth century, drought, infrastructure degradation, and social disorder resulted in abandonment.

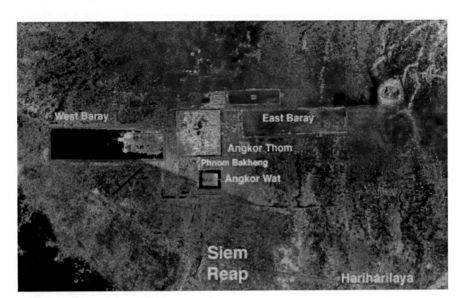

Fig. 10.3 An annotated AIRSAR digital elevation map of Angkor area showing major Barays and ancient infrastructure. The terrain slopes gently from *top to bottom*, which is indicated by the color gradient

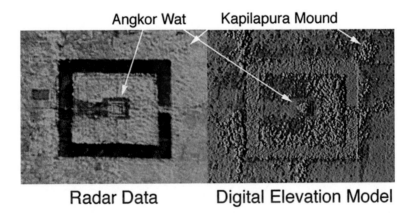

Fig. 10.4 AIRSAR multi-polarization radar image of Angkor Wat (*left*), and high-resolution digital elevation model (*right*). The main temple at Angkor is obvious in both data sets (Note that Kapilarpura Mound is more obvious in the DEM than in the image data)

Multi-polarization radar images of Angkor allowed discrimination between various vegetation types and current land use. The AIRSAR high-resolution digital elevation models provided insight into water use in ancient times and allowed discovery of new structures (see Figs. 9.3 and 9.4).

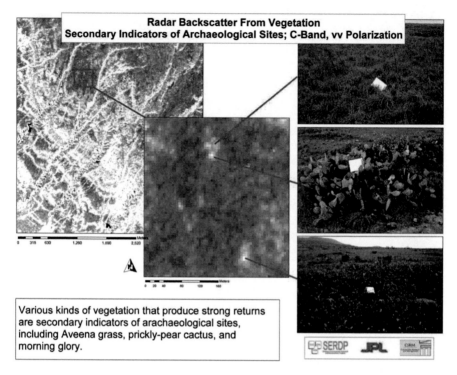

Fig. 10.5 Example C-band imagery from San Clemente Island, collected by AIRSAR in 2004

10.2.3   San Clemente Island, Developing Predictive Archaeological models

Native American archaeology is often more subtle than previous examples and integration of remote sensing into a GIS combined with a traditional archaeological approach is a very practical and powerful approach (Comer and Blom 2007). The U.S. Department of Defense controls more than 24 million acres, for which they are responsible for the protection of cultural remains; much of this land is unsurveyed. A primary goal is simply identifying likely sites so they may be avoided, or efficiently surveying lands needed for other purposes. Accordingly, predictive models using a remote sensing based methodology are extremely useful. San Clemente Island was used as a case study for development of a remote sensing and GIS based predictive model suitable for DoD needs. Controlled since WW II by the Navy, San Clemente Island is about 50 km offshore of San Diego, California. The combination of isolation and naval control has to a large degree minimized modern cultural overprint.

Airborne multi-frequency and multi-polarization radar images by AIRSAR (see AIRSAR 2012) and GEOSAR (see GEOSAR 2012) along with their derived high-resolution digital elevation data sets were acquired for San Clemente Island (see Fig. 10.5) in 2004. In combination with existing data in a GIS, these data were used to develop a preliminary predictive model of potential site locations. Analysis of these data showed that Native American archaeological sites on San Clemente Island were typically radar bright, especially in C-band VV polarization data. Details of this aspect are variable but generally, due to a combination of increased/different vegetation at the sites, are accompanied by subtle soil changes. Very important were high-resolution digital elevation data sets, which revealed that most habitation sites were typically within 200 m of a fresh water supply, on slopes less than 5%. The site sizes were related to the amount of fresh water available, which could be calculated from the digital elevation model. In addition, the high resolution DEM permitted precise ortho-rectification of older air-photos. USGS/SRTM 30 DEMs were not adequate for this task (See Golden and Davenport, this volume). Integration of remote sensing data into an archaeological GIS can be the basis for a predictive model, significantly speeding up preliminary surveys and reducing costs.

10.3   What Is Synthetic Aperture Radar?

A Synthetic Aperture Radar (SAR) transmits a series of microwave pulses towards the desired ground image swath, and then coherently receives the backscattered echoes. The returned signal data is then processed into high-resolution radar imagery. The SAR instrument actively illuminates the ground independent of solar illumination. (See Bamler 2000; ESA Earthnet Online Radar Course III; ESA PolSAR pro tutorial; Tutorial: Fundamentals of Remote Sensing; and Tutorial: Radar Polarimetry for more information.)

10.3.1   SAR Geometry

SAR technology requires that the image swath lie to the side of the nadir path as shown in Fig. 10.6. Because of this side-looking geometry, the initially processed imagery is not projected to the ground, but to an imaginary plane perpendicular to the direction of the wave fronts. As you look across this image swath, the ground resolution will vary with the incidence angle. Conversely, as you look along the flight track, the resolution is constant, and dependent on the interval of the transmitted pulses. This image plane is called the slant plane. Typically the first processing step is to project this slant range image to the ground plane with a uniform resolution.

Because of this imaging geometry, the topography of the earth's surface can affect the appearance of the imagery (see Fig. 10.7). For each transmitted pulse, echoes are collected in increments of distance to the ground (by capturing the data in time increments). Thus, in areas with very steep slopes, the top of a mountain

Fig. 10.6 Synthetic aperture
radar imaging geometry.
The radar transmits pulses of
microwaves and then a short
time later, coherently receives
the backscattered radiation

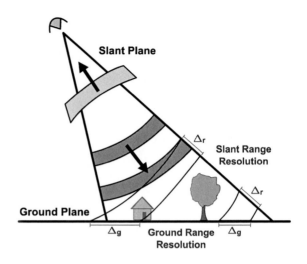

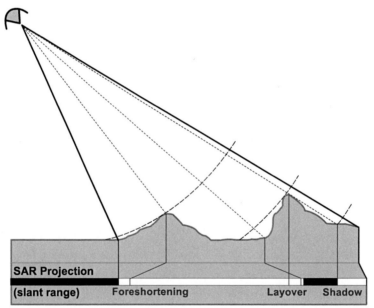

Fig. 10.7 Illustration of layover, shadowing, and foreshortening in radar imagery

may appear to come before the base of the mountain, even though on the earth the
opposite is true. This is called layover. Similarly, even without layover, the top of
the mountain will appear to move closer to the near part of the swath than is the case
on the ground. Called foreshortening, this causes a distortion of the imagery that can
be corrected when the slant range image is projected to the earth's topographic surface. Lastly, if the topography makes it impossible for the radar to receive data from
that range (it is shadowed by the topography), this is called shadowing.

Table 10.1 Common SAR frequency bands, their frequency and wavelength ranges, and their common uses (see Skolnik 1990).

Band name	Frequency range	Wavelength range	Common uses
P	<400 Mhz	1 m+	Penetration, soil moisture, biomass, impacted by other uses (interference)
L	1–2 Ghz	15–30 cm	Vegetation, deformation
S	2–4 Ghz	7.5–15 cm	
C	4–8 Ghz	3.75–7.5 cm	Agriculture, deformation, topography
X	8–12 Ghz	2.5–3.75 cm	High resolution, topography
Ku	12–18 Ghz	1.67–2.5 cm	Ocean and open water
Ka	24–40 Ghz	0.75–1.11 cm	Snow/ice (airborne)
Mm	40–300 Ghz	7.5–1 mm	

10.3.2   SAR Frequency and Polarization

Microwave radars can operate over a large range of frequencies, with the wavelength varying from millimeters to meters. A frequency band consists of a somewhat arbitrary range of frequencies. Different frequency bands require different radar hardware, so usually a radar system can only operate at one frequency band, unless, for example, there are multiple antennas.

Radiation with a wavelength of a millimeter will often interact with objects on the ground very differently than radiation with a wavelength of a meter. For instance, vegetation easily scatters millimeter wavelength radiation, and there is usually little penetration into vegetation (except through gaps) at those wavelengths. On the other hand, meter wavelength radiation can penetrate thick forest canopies, interacting mostly with the ground surface and the trunks of the trees. Often, the frequency of a radar is chosen to take advantage of these frequency dependent scattering properties to optimize its usage. See Table 10.1 for a list of commonly used SAR frequency bands.

As discussed previously, radar can penetrate the soil of hyper-arid environments when using longer wavelengths. The penetration depth is on the order of a several wavelengths in these environments. However, as the soil moisture (among other parameters) increases, the soil becomes increasingly opaque. The backscatter imagery can sometimes reveal features slightly below the surface.

In addition to determining the frequency, the radar also determines the polarization. The polarization is the direction of the electromagnetic oscillations perpendicular to the direction of propagation. For example, horizontal polarization means that the electromagnetic oscillations are lying in a horizontal plane perpendicular to the direction of propagation. Usually SAR antennas transmit and receive horizontal and/or vertical polarizations, but other polarizations, such as circular polarizations are possible. A fully polarimetric radar is one that can transmit and receive two orthogonal polarizations (such as horizontal and vertical). A single polarization radar is one that transmits and receives a single polarization, while a dual polarization radar is one that can either transmit or receive two polarizations.

Polarization is important because the objects on the ground scattering the radiation can change the polarization. For example, if an L-band radar transmits horizontal polarization towards a forested area, after the multiple scattering that occurs within the forest canopy, we expect to have a significant signal returned that is now vertically polarized. With fully polarimetric radar, it is possible to determine how much the ground target changed the polarization. In cases where fully polarimetric SAR data is not available, dual or single polarization data will still be sensitive to the polarization properties of the scatterers on the ground, but the exact change in polarization may be difficult to quantify.

10.3.3   SAR Image Brightness

The brightness of the radar image is dependent on many properties of the objects that are scattering the radiation, but the scattering interaction that occurs between the electromagnetic waves and these surface objects often dominate the appearance of the image. The configuration of the radar (frequency and polarization) will additionally determine the nature of the scattering.

There are three *primary* ways that radar waves scatter off the ground: rough surface scatter, diffuse scatter, and double bounce scatter (see Fig. 10.8).

Surface Scatter   For the rough surface scatter case, most of the transmitted radiation is scattered away from the direction of the radar, but depending on roughness, some of the energy is reflected off the surface back in the direction of the radar.

Diffuse Scatter   For the diffuse scattering case, the electromagnetic waves scatter off multiple surfaces within a layer of objects (such as vegetation), and the resulting energy is reflected uniformly in all directions.

Double Bounce   In the case of double bounce scattering, the electromagnetic waves reflect off of two perpendicular surfaces (such as a wall lying across a flat surface), and reflect most of the energy back to the radar.

These different scattering mechanisms result in three very different brightnesses in the imagery, from dark (rough surfaces), to medium brightness (diffuse), to bright (double bounce). The scattering mechanism depends also on the orientation angles and incidence angles, as well as the transmitted polarization of the radar. If a wall is not parallel to the flight track of the radar, a double bounce reflection might not occur. At some incidence angles, rough surface scatter can become quite bright, such as C-band imagery over wind- roughened water. Factors that affect the dielectric constant, such as moisture content and freeze/thaw state, can also affect the radar image brightness.

Figure 10.9 shows two examples of double bounce reflections. One is caused by power poles aligned to the flight track, and one is caused by inundated vegetation, in which the smooth water below the emergent vegetation results in a strong double bounce signature off of tree trunks.

Fig. 10.8 Dominant radar
scattering mechanisms

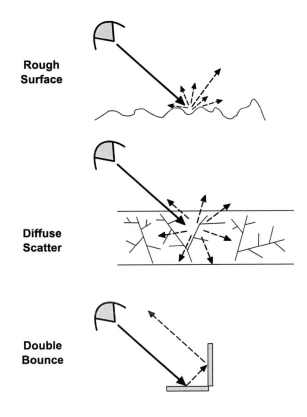

**Rough
Surface**

**Diffuse
Scatter**

**Double
Bounce**

Topography also plays a major role in image brightness, but some effects can be corrected if the topography is well known. For instance, if a hill slope is facing away from the radar, the area of the ground illuminated by the radar will be smaller than if there was no hill, and therefore the brightness on these slopes will be less than that of flat terrain. In addition, different targets on the ground will interact differently with the radar waves at different incidence angles. Therefore, there is a dependence on incidence angle that is dependent on the properties of the objects being illuminated. While the latter cannot be easily corrected, the former can be (if the topography is well known). Figure 10.10 shows a L-band SAR image before and after the image brightness has been corrected for slope.

Since the scattering mechanism can indicate the structural characteristics of the objects within the image swath, such as urban structures (double bounce), forests (diffuse scatter), or bare ground (rough surface scatter), tools have been developed to automatically classify the data based on the polarimetric characteristics of the data. These tools will examine the data, and estimate for each pixel in the image the scattering mechanism (if fully polarimetric data), and may also attempt to automatically classify the terrain type from the scattering mechanism.

SAR images can be displayed as black and white images, where the image brightness corresponds to the backscattered radar brightness; or displayed as color

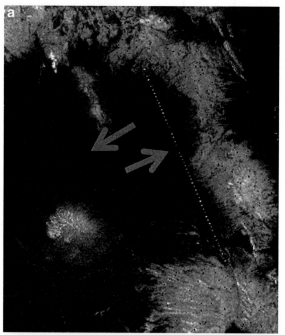

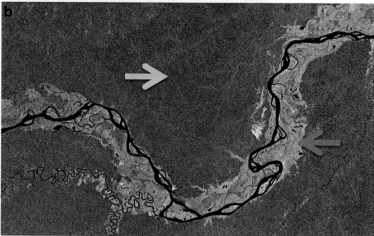

Fig. 10.9 ALOS PALSAR L-band radar imagery. (**a**) Eureka, Nevada, USA, power poles, one example indicated by *blue arrow*, are aligned with the flight track, and show up as bright points in the imagery (© JAXA/METI). (**b**) Flooded forest along the Amazon River at the borders of Brazil, Peru, and Colombia, shows up as bright regions in this L-band HH image (for example, where *orange arrow* is pointing) (© JAXA/METI). The dark areas in (**a**) illustrate the appearance of dark surface scatter (for example, where *red arrow* is pointing), while the non-flooded areas in (**b**) illustrate the appearance of diffuse scattering (for example, *yellow arrow*)

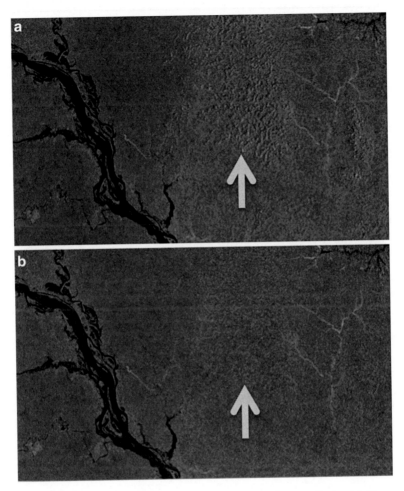

Fig. 10.10 ALOS PALSAR L-band HH polarization image, Amazon Basin. (**a**) Uncorrected L-band HH image with subtle topographic slope features in the middle of the image (*yellow arrow*). (**b**) Corrected for terrain slope (slope effects minimized at *yellow arrow*) (©JAXA/METI)

images, where either different polarizations or different frequencies are assigned colors and then combined.

The values of each pixel in a SAR image file are usually calibrated such that the pixel value is in units of "radar cross section", which is proportional to the ratio of the power density impinging the ground to that received back at the radar. Usually, these values are recorded in the SAR data file as integers or floating point numbers, which can be converted to radar cross section in decibels (dB) (a logarithmic scale) through a calibration factor often provided with the data. Since there is a strong preference for most of the energy to reflect away from the radar due to the side-looking geometry, the observed ratio is almost always less than one. The dB values are therefore usually negative, typically ranging from −5 dB (bright) to −20 dB or less (dark).

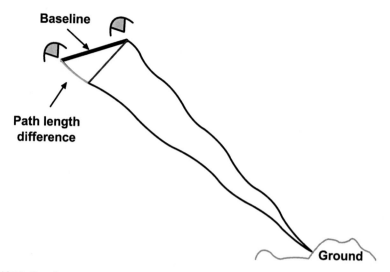

Fig. 10.11 Interferometric SAR requires at least two observations of the same ground location, from approximately the same location in space. These observations do not have to occur simultaneously. Shown is a non-zero baseline

10.3.4 Interferometric SAR

Interferometic Synthetic Aperture Radar (InSAR) is a technique that requires multiple acquisitions of data (see Fig. 10.11). These data can be acquired at the same time (single-pass InSAR), and/or at different times (repeat-pass InSAR). The data are either acquired by the instrument from nearly the same position (zero baseline) or from an offset position (non-zero baseline).

Topography or surface deformation can be estimated by interfering the two complex images from the two acquisitions, and producing an intermediate product called an interferogram. Another intermediate product called the coherence image indicates how statistically similar the two complex images are. From the interferogram, as well as knowledge of the baseline between observations and parameters of the radar, the interferogram can be interpreted and used to estimate ground topography or topographic change for each pixel in the image. InSAR data are typically displayed as a combination of the SAR backscatter image overlaid with colors indicating displacement or topography (or by the related phase change, which may be a combination of the two).

Data acquired from a zero baseline are very sensitive to change. For instance, if the topography of the surface changes by even a fraction of a wavelength between acquisitions, this can be detected. Zero baseline data are only useful over land when the acquisitions are repeat-pass, when there is a time interval between the observations. Non-zero baseline data are useful for estimating topography, as well as being sensitive to structure within the topography (such as vegetation). It is

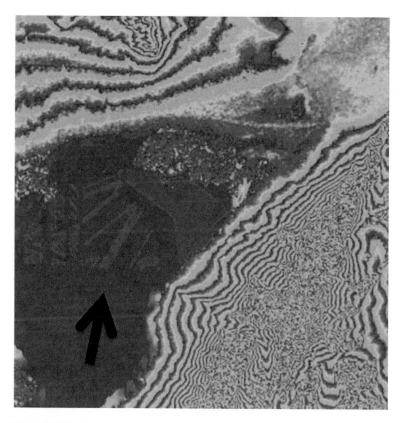

Fig. 10.12 AIRSAR C-band single pass INSAR digital elevation model. Colors wrap around from *blue* to *blue* every 10 m in height. Slight man-made features are visible in the flat area (see *black arrow*)

most useful when these acquisitions are single-pass, so that temporal changes between observations don't contaminate the data. In practice, most acquisitions that are repeat-pass are also non-zero baseline, as zero baselines can be difficult to accomplish in repeat-pass mode. Since non-zero baselines are sensitive to both topography and topographic change, the additional complexity of topography must be compensated if topographic change information is desired. Topography can be difficult to measure from non-zero baseline repeat-pass data, since changes in the atmosphere and ionosphere also impact the interferogram, and because the baseline (and therefore the sensitivity to topography) typically varies with time.

Figure 10.12 shows results from a portion of an AIRSAR C-band Digital Elevation Model (DEM), similar in nature to the near-global NASA C-band Shuttle Radar Topography Mission (SRTM) derived DEM (SRTM 2012) measured by single-pass non-zero baseline interferometry. As can be seen, even very slight differences in relative topography can be discerned in the flatter portions of the image.

Figure 10.13 shows deformation (change in topography) between two repeat-pass INSAR observations by the NASA L-band airborne SAR called UAVSAR

Fig. 10.13 This figure shows the phase change in the radar interferogram due to an earthquake that occurred between two observations by UAVSAR, an L-band airborne repeat pass InSAR system. The color "fringes" indicate the deformation (in the line of site direction to the radar) that occurred. Agricultural areas (*white arrow*) often show a signature in repeat pass interferograms as well (SanAnd_26501_09083-010_10028-000_0174d_s01_L090_01, UAVSAR 2012)

(see UAVSAR 2012), in which the observations straddle in time a major earthquake in the region. Here, the colors are wrapping every 12 cm and are indicating small changes in topography caused by the earthquake. Figure 10.14 shows a UAVSAR repeat-pass interferogram showing subsidence on the order of tens of centimeters.

10.4   Application of SAR Data

There are many software packages available that make it easy to use SAR data, some of them open source and free (see Table 10.2). There are generally three objectives of the software: the display, classification, and manipulation of the radar data. Display includes reading the file format (which can be non-standard), and enabling display of the data, either internally within the software or in another software package. Classification encompasses not only reading in the data, but identifying features in the data revealed in the polarimetric characteristics of the data, or more simple classifications based simply on the image brightness. Manipulation of the data covers a wide variety of procedures, from processing raw signal data into imagery to calibration to re-projection of the data.

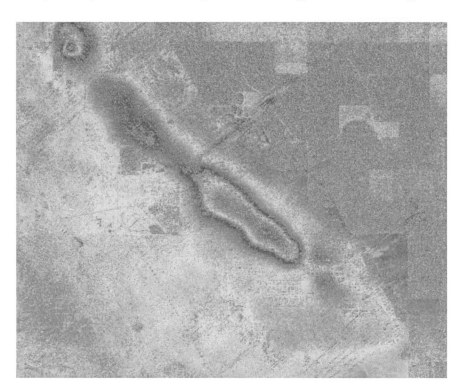

Fig. 10.14 An L-band HH UAVSAR repeat pass (about 1 year apart) InSAR image of the South Belridge oil field area in California that shows subsidence of about 20 cm (SanAnd_05006_09084-009_10076-011_0367d_s01_L090_01, UAVSAR 2012)

Table 10.2 Some useful free SAR software as of May 2012

Name	Purpose	URL for download and more information
ASF Mapready	Display, classification, manipulation	http://www.asf.alaska.edu/downloads/software_tools
ESA Polsarpro	Display, classification	http://earth.eo.esa.int/polsarpro/
ESA NEST SAR Toolbox	Display, manipulation	http://nest.array.ca/web/nest
Roipac	Manipulation	http://www.roipac.org/ROI_PAC

There are many available SAR data sets, both spaceborne and airborne. Some of these data sets can be obtained at little or no charge, and some can be quite expensive; the cost will depend on the source of the data. Some data can be obtained through intermediaries that distribute data, perhaps with some value added. See Table 10.3 for a sampling of easily available data (though certainly not

Table 10.3 Common SAR data sources as of May 2012

Agency	Data	URL for more information
NASA and ASF	JERS, ERS, Radarsat, AIRSAR, UAVSAR, PALSAR	http://www.asf.alaska.edu http://uavsar.jpl.nasa.gov http://airsar.jpl.nasa.gov
USGS	SRTM and SIR-C	http://earthexplorer.usgs.gov
CSA	Radarsat I, II	http://gs.mdacorporation.com/
ESA	ERS, Envisat	http://earth.esa.int/ers/ http://envisat.esa.int/instruments/asar
JAXA	JERS, ALOS	http://en.alos-pasco.com/ http://www.eorc.jaxa.jp/en/about/ distribution
Infoterra	TerraSAR X, Tandem X	http://www.infoterra.de/

a complete list, as data availability changes with the addition and subtraction of various SAR resources).

The radar data come in a variety of data formats and processing levels. Sometimes the data are available in the slant range projection in a customized format, sometimes the imagery will be projected to the Earth's ellipsoid, and sometimes the data will be available projected to the Earth's topography where the image brightness has been corrected for some topographic slope effects. The software in Table 10.2 may or may not support all formats and projections; selecting the correct processing level depending on the desired usage and the software should be an important consideration in deciding which data to acquire.

10.5 Conclusions

Synthetic Aperture Radar is a powerful technique for remotely imaging the earth that has been shown to have applications in archaeology. Current sensors such as TandemX, an X-band single pass InSAR system designed for high-resolution topographic mapping, and future planned sensors such as DESDYnI, an L-band repeat pass InSAR system for mapping crustal deformation, the cryosphere, and ecosystem structure, will be mapping regions worldwide. The archaeological community, armed with information on how to use SAR, will undoubtedly benefit from exploiting these data as well as other remote sensing data.

Acknowledgments We thank and acknowledge Sarah Flores of JPL for creating some of the illustrations in this chapter. We thank Craig Dobson of NASA HQ for his support. This work was performed at the Jet Propulsion Laboratory, California Institute of Technology, under contract with the National Aeronautics and Space Administration.

References

AIRSAR (n.d.). Airborne synthetic aperture radar. http://airsar.jpl.nasa.gov. Accessed 9 May 2012.

Bamler, R. (2000). Principles of synthetic aperture radar. *Surveys in Geophysics, 21*(2/3), 147–157. Retrieved from https://ejournal.csiro.au/cgi-bin/sciserv.pl?collection=journals&journal=0169 3298&issue=v21i2-3&article=147_posar.

Blom, R. G., Crippen, R. E., & Elachi, C. (1984). Detection of subsurface features in Seasat radar images of Means Valley, Mojave Desert, California. *Geology, 12*, 346–349.

Comer, D., & Blom, R. (2007). Detection and identification of archaeological sites and features using synthetic aperture radar (SAR) data collected from airborne platforms. In J. R. Wiseman & F. El-Baz (Eds.), *Remote sensing in archaeology* (pp. 103–136). New York: Springer Science and Business Media.

Elachi, C., Roth, L. E., & Schaber, G. G. (1984). Spaceborne radar subsurface imaging in hyper-arid regions. *IEEE Transactions on Geoscience and Remote Sensing, GE-22*, 383–388.

ESA Earthnet Online Radar Course III (n.d.). http://earth.esa.int/applications/data_util/SARDOCS/ spaceborne/Radar_Courses/Radar_Course_III/. Accessed 9 May 2012.

ESA PolSAR pro tutorial. (n.d.). http://earth.esa.int/polsarpro/tutorial.html. Accessed 9 May 2012.

Evans, D. L., Plaut, J. J., & Stofan, E. R. (1997). Overview of the spaceborne imaging radar-C/X-band synthetic aperture radar (SIR-C/X-SAR) missions. *Remote Sensing of Environment, 59*(2), 135–140. doi:10.1016/S0034-4257(96)00152-6. ISSN 0034–4257.

Farr, T. G., Elachi, C., Hartl, P., & Chowdhury, K. (1986). Microwave penetration and attenuation in desert soil: A field experiment with the shuttle imaging radar. *IEEE Transactions on Geoscience and Remote Sensing, GE-24*(4), 590–594.

GEOSAR – Mapping the impossible with dual-band IFSAR (n.d.). http://www.geosar.com. Accessed 9 May 2012.

Greater Angkor Project. (n.d.). http://acl.arts.usyd.edu.au/angkor/gap/. Accessed 9 May 2012.

McCauley, J. F., Schaber, G. G., Breed, C. S., Grolier, M. J., Haynes, C. V., Issawi, B., Elachi, B., & Blom, R. (1982). Subsurface valleys and geoarchaeology of the eastern Sahara revealed by shuttle radar. *Science, 218*, 1004–1020.

Moore, E., Freeman, A., & Hensley, S. (2007). Spaceborne and airborne radar at Angkor: Introducing new technology to the ancient site. In J. R. Wiseman & F. El-Baz (Eds.), *Remote sensing in archaeology* (pp. 185–216). New York: Springer Science and Business Media.

Paillou, P., Schuster, M., Tooth, S., Farr, T., Rosenqvist, A., Lopez, S., & Malezieux, J. (2009). Mapping of a major paleodrainage system in eastern Libya using orbital imaging radar: The Kufrah River. *Earth and Planetary Science Letters, 277*(3–4), 327–333. ISSN 0012-821X, 10.1016/j. epsl.2008.10.029. (http://www.sciencedirect.com/science/article/pii/S0012821X08006924)

Skolnik, Merrill I. (1990). Radar Handbook (2nd Edition). McGraw-Hill. Online version available at: http://www.knovel.com/web/portal/browse/display?_EXT_KNOVEL_DISPLAY_bookid= 701&VerticalID=0

SRTM – Shuttle Radar Topography Mission. (N.d.). http://www.jpl.nasa.gov/srtm. Accessed 9 May 2012.

Tutorial: Fundamentals of Remote Sensing. (n.d.). *Natural resources Canada*, Canada Centre for remote sensing. http://www.nrcan.gc.ca/earth-sciences/geography-boundary/remote-sensing/ fundamentals/1430. Accessed 9 May 2012.

Tutorial: Radar Polarimetry. (n.d.). *Natural resources Canada*, Canada Centre for Remote Sensing. http://www.nrcan.gc.ca/earth-sciences/geography-boundary/remote-sensing/radar/1893. Accessed 9 May 2012.

UAVSAR – Uninhabited Airborne Synthetic Aperture Radar. (N.d.). http://uavsar.jpl.nasa.gov. Accessed 9 May 2012.

Chapter 11
The Use of Multispectral Imagery and Airborne Synthetic Aperture Radar for the Detection of Archaeological Sites and Features in the Western Maya Wetlands of Chunchucmil, Yucatan, Mexico

David R. Hixson

Abstract This case study presents the results of an extensive remote sensing survey testing the capabilities of multispectral and Synthetic Aperture Radar data to detect ancient Maya settlements in the seasonally inundated near-coastal region of northwest Yucatan. The results are compared to similar recent studies published by researchers working in the southern Maya lowlands. It is concluded that seasonal and regional variation across the Maya area precludes universal application of a singular remote sensing technique using these and similar platforms. Instead, specific climatic, seasonal, and physiographic context must be considered in the selection of a remote sensing data set. For multispectral data, spectral contrast, spectral resolution, and temporal resolution appear to be more critical for site detection than spatial resolution. For the AirSAR polarimetric data, it is concluded that the L-band (rather than the highly anticipated P-band) resulted in a better correlation between positive returns and monumental architecture within the forest canopy. Yet, the C-band AirSAR DEM appears to have the broadest application in Maya archaeological survey, especially when combined with other remote sensing data sets.

Keywords Maya • Remote sensing • Landsat • Multispectral • Synthetic aperture radar • AirSAR • Chunchucmil • Vegetation • Seasonal wetlands

11.1 Introduction

Archaeological remote sensing has a long (yet mixed) history in the Maya area, with many pioneering efforts advocating the wide applicability of new technologies, often followed by published studies critiquing the practical applications of these same technologies (e.g., Adams 1982; Pope and Dahlin 1989). With each additional study, researchers refine our knowledge of both the capabilities and limitations of available remote-sensing platforms for the field of archaeology. The results of the

D.C. Comer and M.J. Harrower, *Mapping Archaeological Landscapes from Space*,
SpringerBriefs in Archaeology, DOI 10.1007/978-1-4614-6074-9_11,
© Springer Science+Business Media New York 2013

following case study from Chunchucmil, Yucatan, when placed within the context of our seemingly endless effort to peer below the vegetation blanketing southern Mexico and northern Central America, suggest why some projects seem to have greater success than others when utilizing the same technology.

11.2   Multispectral Remote Sensing of the Maya Area

Early use of multispectral Landsat data to directly detect ancient Maya features initially resulted in a better understanding of the road systems connecting ancient capitals to their secondary centers. Folan et al. (1995: 277–283) found that causeways could be detected as linear peaks in the infrared bands, indicating lines of vigorous plant growth through low swampy areas. The authors also found that those pixels combining to form indications of "dryness" (using a tasseled cap transformation [Lillesand and Kiefer 2000: 523]) could differentiate causeways from inundated terrain (1995: 279). Subsequent studies by NASA archaeologist Tom Sever (1998) verified the utility of this approach for the detection of ancient road beds. But specific spectral reflectance values directly indicating the presence of habitation sites beneath the forest canopy remained elusive.

A recent publication by Saturno et al. (2007) briefly discussed the limitations of Landsat imagery for the detection of ancient Maya archaeological sites, and indicated that the sensor's spatial resolution (30 m multispectral, 15 m panchromatic) was its greatest constraint. Instead they examined the potential for Very High Resolution (VHR) multispectral imagery (IKONOS). The IKONOS platform produces 4 m multispectral (3 bands visible and 1 near-infrared) and 1 m panchromatic data sets. In the pan-sharpened multispectral data, the authors observed a spectral reflectance indicating canopy stress over known Maya centers in the San Bartolo region. Upon further investigation, this was found to correlate with previously undocumented sites. The researchers concluded that VHR multispectral imaging was successful in predicting the locations of archaeological sites, from large centers to small tertiary settlements, and that the imagery "had the potential completely to revolutionize archaeological survey in the tropics by dramatically reducing the time involved in systematically covering vast areas" (2007: 149).

Yet, an attempt by Garrison et al. (2008) to replicate this process in other areas of the Maya lowlands demonstrated that archaeological remote sensing has not yet found that surveyor's panacea envisioned by Saturno. Garrison and his team found that the IKONOS "vegetation signature" observed by Saturno is weak to non-existent over other major Maya centers in neighboring regions (e.g., Ceibal). Meanwhile a spectral response even stronger than that noted by Saturno was identified by Garrison as a false positive reading (only two platforms where the satellite imagery indicated a major center in the Sierra de Lacandon).

The conclusion of Garrison and his colleagues was that "local climate, geology, hydrology, topography, pedology, and vegetation" differs significantly from region to region, making broad application of Saturno's observations at San Bartolo problematic (Garrison et al. 2008: 2770). Garrison also notes the problems inherent in seasonal climatic variations when attempting to compare reflectance values across images taken at different times of the year.

11.3   Multispectral Remote Sensing of the Chunchucmil Region

The site of Chunchucmil is located in the northwest corner of the Yucatan peninsula, Mexico. First identified in the mid-1970s by the *Atlas Arqueólogico del Estado de Yucatán* (Garza and Kurjack 1980), this metropolis is comprised of several 1,000 house mounds and dozens of temples jutting out of the relatively planar landscape that is the Yucatecan karstic plains physiographic region.

The lack of topographic relief, combined with the lowest rainfall total in all the Maya lowlands, makes this scrub forest ideal for the application of remote-sensing techniques for archaeology, especially when compared to the lush tropical jungles of the southern Maya realm. Chunchucmil was readily detected by the Atlas project using high-resolution stereoscopic aerial photographs. From these photos, they were able to describe the site center and map a portion of the residential core (Garza and Kurjack 1980; Vlcek 1978; Vlcek et al. 1978) using limited field verification.

Decades later, our Pakbeh Regional Survey subproject (1999–2005) decided to build upon the successes of the Atlas project by applying multispectral imaging to our study of Chunchucmil's overall size, layout, and regional settlement pattern. We began by examining nearly three decades of multispectral Landsat data, looking for images that clearly differentiated the urban core of Chunchucmil from its natural surroundings.

Since seasonality dramatically affects the surface configuration of the region, we acquired both dry-season and rainy-season images for all decades of the Landsat mission. We found that multispectral images following early heavy rains were infinitely more valuable than those images taken during the height of a drought. In July of 1999, when a tropical storm stalled over this corner of Yucatan, our project was present to note the seasonal wetlands encroaching upon the sites center, with only the archaeological mounds and other cultural features protruding from the temporary inundation. The water quickly receded, but the effects on regional vegetation patterns were pronounced.

By the end of July, the secondary growth covering any partially cleared archaeological site burst into life. Even the most modest residential mounds were composed of anthropogenic soils that supported green leafy vegetation while the surrounding terrain was filled with either still-receding flood waters or recovering low-lying grasslands. Bands 4 and 5 of Landsat 7 (near- and mid- infrared) highlighted not only the urban core of Chunchucmil, but also several large secondary sites previously noted by the Atlas project (Fig. 11.1).

In addition to using the infrared bands, we were able to generate new bands using the same tasseled cap transformation that was so beneficial to previous Maya scholars. The algorithm generates three new bands characterized as "brightness,

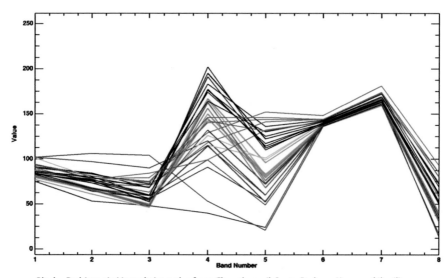

Black: Prehispanic Mounds (samples from Chunchucmil, Santa Barbara, Kum, and Ikmil)
Green: Tree Cover (from "Yax Hom" through "Monte Alto" to "Monte Bajo")
Red: Bare Soil (Kancab flats with grass cover)
Blue: Water (including coastal waters, cenotes, flooded savanna, and flooded forest)

Fig. 11.1 Spectral plot of selected pixels from Landsat 7 ETM + imagery of NW Yucatan, July 1999, showing the greatest spectral contrast in the near- and mid-infrared bands (4 and 5)

greenness, and wetness" (Lillesand and Kiefer 2000: 523). All of the archaeological sites that could be detected in the multispectral imagery (Fig. 11.2) share the characteristics of being bright (broken limestone surfaces), green (early lush vegetation), and relatively dry, when compared with surrounding natural features such as mature canopy (blue-green in this imagery), bare rocky soil or barren fields of dead grass (red), or inundated *bajos* (deep blue).

We then used the Landsat imagery to estimate the size of ancient Chunchucmil. It was concluded that the most densely settled portion of Chunchucmil covered an area of roughly 25 sq km, with a more sparsely populated suburban zone that increased this number to nearly 64 sq km (Hixson 2005). Years later, pedestrian mapping transects revealed that the actual settlement density of Chunchucmil was consistent with the multispectral remote sensing estimates (Hutson et al. 2008).

We next established six targets within the western seasonal wetlands to test the sensor's capability of detecting secondary or tertiary settlements beyond the urban core of Chunchucmil. All six resulted in the discovery of previously undocumented settlements, from secondary ceremonial centers to rural farmsteads and seasonal camps (Hixson 2011).

The lessons learned during the multispectral remote-sensing surveys of Chunchucmil revolve around seasonality and synchronic on-the-ground observations. In an area where tidal fluctuations, local rainfall, surface drainage, and minor

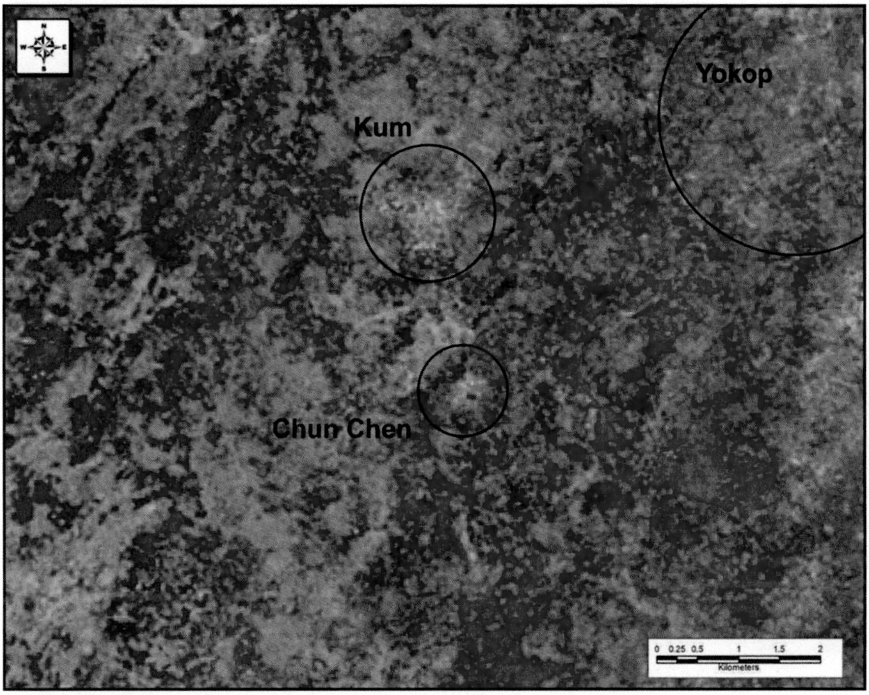

Fig. 11.2 Tasseled cap transformation of Landsat 7 ETM + imagery from July 1999, highlighting the western regional sites of Kum, Chun Chen, and Yokop. *Yellow* = Bright, Green, and Dry

topographic rises dramatically affect the surface configuration (and potential ancient settlements), the selection of remotely sensed imagery based upon specific and observable seasonal conditions appears to be crucial.

Only one Landsat image out of dozens proved to clearly differentiate "site" from "non-site". This supports Garrison's preliminary conclusions regarding the importance of regional and temporal variation in climate and ecology (Garrison et al. 2008). Our research at Chunchucmil suggests that one must find the imagery that precisely captures the time of year when vegetation is under the *proper* stress to highlight different surfaces through elevated spectral contrast. If the region is uniformly dry, the result will be "dry rocky mounds" against "dry rocky ground". If the region is uniformly wet, the imagery will be blanketed by a drape of impenetrable greenery.

The Chunchucmil data also indicates that Saturno's focus upon VHR (very high *spatial* resolution) imagery is misplaced. The 30 m Landsat TM data was more than sufficient to differentiate site from non-site under the proper regional and seasonal conditions. In fact, the addition of the mid-infrared band (LS-band 5 – not available on IKONOS) was critical to the application of the tasseled cap transformation. Even without the tasseled cap transformation, archaeological sites around Chunchucmil were clearly indicated in a contrast stretch of bands 5,4,2 of the Landsat data (Fig. 11.3). But most importantly, Landsat's vast archive of images extending back

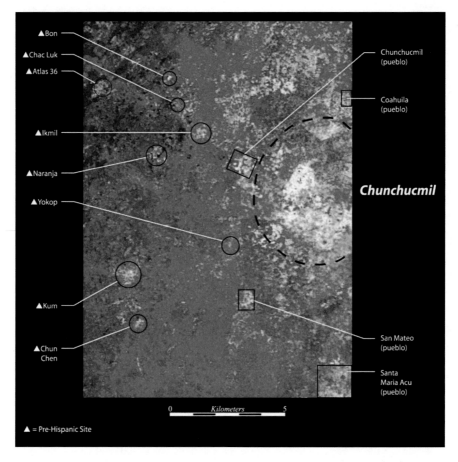

Fig. 11.3 Landsat 7 ETM + image (bands 5,4,2) from July of 1999, histogram equalization stretch, highlighting Chunchucmil and its western hinterland. Peaks in all three bands (mid-infrared, near-infrared, and visible green) result in white pixel clusters over archaeological sites and modern villages

to the 1980s, with regular and repeated passes over the entire region, provided greater *temporal* resolution, allowing for the precise selection of an image based upon specific seasonal conditions.

11.4 Synthetic Aperture Radar Survey of the Chunchucmil Region

The Chunchucmil project, along with over a dozen other programs, also participated in the "AirSAR Mesoamerica" campaign (Blom et al. 2003). Collectively, we decided that the experimental use of the P-band would be most beneficial, as it has

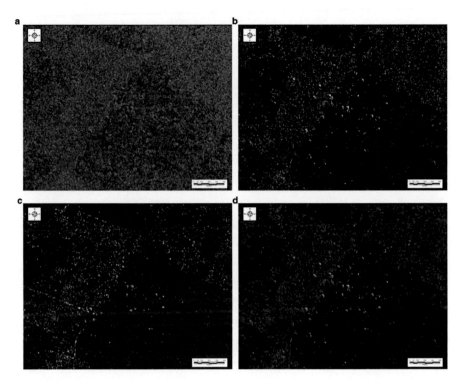

Fig. 11.4 AirSAR data of the site center of Chunchucmil. (**a**) C-vv; (**b**) L-hv; (**c**) P-hv; (**d**) L-hh, L-hv, and P-hv (in RGB color space)

the longest wavelength and therefore the greatest likelihood of penetrating the dense canopy of southern Mesoamerica for the detection of archaeological features (Pope et al. 1994). The instrument was flown in "TOPSAR" mode, generating both fully polarimetric imaging data and a C-band DEM (see Chapman and Blom, this volume). The AirSAR instrument flew over the Chunchucmil region in four swaths measuring approximately 50 km long by 10 km wide in March of 2004. The resulting data covers roughly 2,000 sq km.

The AirSAR Mesoamerica campaign originally had high hopes for the P-band. We were finally going to "pierce" the forest canopy and receive backscatter signals from the actual ground surface. At the AirSAR Mesoamerica roundtable, held 1 year after deployment, none of the participants (from Tikal to Chunchucmil) achieved this outcome. Since that time, only one publication has offered a newly discovered site based upon the AirSAR Mesoamerica data (Garrison et al. 2011; using the C-band DEM rather than the P-band imaging data). The Chunchucmil project, however, appears to have provided the best results to explain the failings of the P-band to live up to our hopes.

Figure 11.4 illustrates the differences between the C-band, L-band and P-band results for the site center of Chunchucmil. Note that at the time of data acquisition

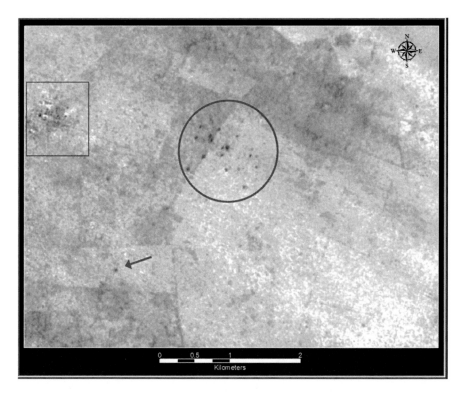

Fig. 11.5 AirSAR DEM (color inversed) showing the tight cluster of pyramidal mounds within the 1 sq-km site center of Chunchucmil, with one outlier known as "the south group." The modern village of Chunchucmil appears in the *upper-left*

the eastern half of the site center was relatively denuded of vegetation, while the western half was in modest forest with a relatively dry 15-to-20 m-tall canopy. The C-, L- and P-bands all detected the tallest pyramids, even within the forested western section of the site center, however the C-band (the shortest wavelength, at 6 cm) and the P-band (the longest wavelength, at 68 cm) also provided false readings that could have been confused for architecture without previously ground-truthed data sets.

The L-band (at 24 cm) provided a near perfect correlation between positive readings and monumental architecture within both the cleared and the forested portions of the site. The longer P-band *did* pierce the canopy, but also provided direct or double-bounce backscatter readings for various natural structures beneath the canopy. Meanwhile, the C-band provided an abundance of returns on smaller natural features of the upper canopy or lower undergrowth. The L-band provided a randomly oriented scatter off the general backdrop of the dry forest canopy, thus allowing the remote-sensing software to highlight where there were sharp anomalies in the canopy itself (backscatter off of the pyramids or the largest trees that grow upon them) without being confused by naturally occurring surfaces beneath the canopy or at its crest.

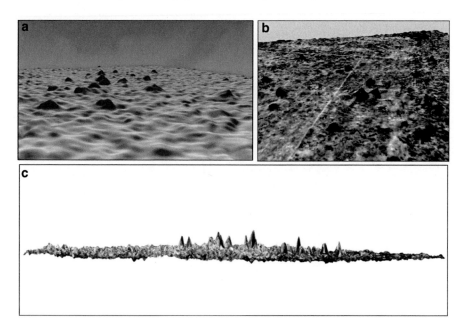

Fig. 11.6 AirSAR digital elevation model of Chunchucmil (**a**) AirSAR DEM perspective view of the site center facing east; (**b**) AirSAR DEM with an image drape of INEGI orthorectified aerial photos, perspective view of the site center facing southeast; (**c**) AirSAR DEM profile view of the site center, Chunchucmil. Vertical scale amplified

Figure 11.5 illustrates the AirSAR C-band DEM of Chunchucmil's site center. The largest architecture is tightly clustered within the central 1 sq km, matching all published estimates for the urban core. Other smaller pyramids can be found in the imagery, but it is clear that Chunchucmil had a tightly clustered nucleus of monumental architecture. A perspective view of the DEM illustrates the ability of the sensor to detect mounded architecture within both cleared fields and secondary forest (Fig. 11.6a). By merging this DEM with aerial photographs (Fig. 11.6b), not only do the pyramids "pop" out of the imagery, but so do the quadrangular arrangements associated with each pyramid. When viewing the DEM in profile (Fig. 11.6c) the monumental architecture of Chunchucmil appears as a normalized curve, highlighting the site's tendency towards a centralized settlement pattern even within the site core. This DEM also led our survey crew to discover a previously undocumented archaeological site outside of the town of Dzidzibalchi, Yucatan, where (in a situation similar to Garrison et al. 2011) large natural hills had been heavily modified to act as foundations for architecture – causing this secondary settlement to appear like a large built city in the AirSAR DEM. The DEM was also used in concert with the multispectral imagery discussed above to locate a Middle Preclassic ballcourt site far into the wetlands west of Chunchucmil (Hixson 2011).

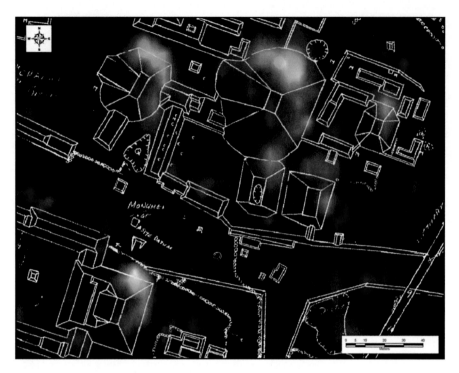

Fig. 11.7 AirSAR bands L-hh, L-hv, and P-hv (in RGB color space) of the pyramidal structures surrounding the site center datum, overlaid with the georeferenced hand-drawn field maps

Another important observation for the detection of cultural remains using AirSAR is the benefit of polarization and the prevalence of double-bounce returns. The bands that returned the best data for the detection of ancient features of Chunchucmil's site center were those that were sent with horizontal polarization and received through a vertical filter (specifically L-hv). Figure 11.4d provides a false color composite of the site center using three bands of the AirSAR data (L-hh, L-hv, and P-hv) and Fig. 11.7 overlays our hand-drawn map of the site center with those same three bands. The strongest returns from the L-hv band (in green) cluster at the northeast corners of each pyramid's basal platform. This highlights two factors influencing the results. First, the look angle was originating from the northeast (peering southwest). Thus, the northeast sides of structures produced the strongest returns, while the southwest sides of structures remained dark (within a "shadow effect"). This is also reflected in the DEM (Fig. 11.6a) where elevation details are most accurate on the northeast slopes of structures, while the SAR "shadow effect" creates false tailings on the southwestern slopes. Second, the highest returns in the cross-polarized L-band data from the bases of pyramids indicate a strong "double-bounce" scenario (see Chapman and Blom, this volume).

In the end, the AirSAR platform has proved an excellent resource for locating a major ancient Maya site center within the semi-arid scrub forests of northwest

Yucatan. While the Pakbeh project already knew the locations of all major structures within the site center of Chunchucmil prior to the deployment of the platform, this test of the sensor allowed a more synoptic view of the region. Since the AirSAR swaths covered approximately 2,000 sq km, a thorough review of the DEM suggests that no other major Maya centers exist within this remote-sensing survey area. Nearly all monumental architecture resides within a centralized 1 sq km (the site center of Chunchucmil). This places Chunchucmil at the apex of its regional settlement hierarchy, potentially dominating a much larger area than many proposed pre-Columbian polities. Plus, AirSAR has demonstrated the lack of any first- or second-tier competitors which are relatively common in other lowland Maya regional hierarchies.

11.5   Conclusions

Within the seasonal wetlands of the northern Maya lowlands, spectral contrast is at its height following the first heavy rains of the season. While this meteorological situation cannot be predicted, it is ideally observed through remote sensing platforms that capture images over a repeated interval (such as searchable Landsat archives). Multispectral platforms with a high temporal resolution result in greater likelihood of useful returns than high spatial resolution scenes from a tasked platform. Furthermore, we conclude that high spatial resolution is not necessary for site detection. Rather, spectral resolution and heightened spectral contrast must be emphasized in the selection of remote-sensing platforms and images.

Similarly, while the AirSAR platform produced a dramatic image of the site center of Chunchucmil, its utility was limited beyond the monumental core due to the lack of contrast between seasonally dry *bajos* and perennially dry uplands. Synthetic aperture radar excels at differentiating inundated surfaces in seasonal wetlands (Pope et al. 1994, 2001). But the data from the height of the dry season at Chunchucmil showed that structural contrast is reduced under extremely arid conditions. It is the recommendation of this project that future SAR missions over seasonally inundated terrain fly during the rainy season.

Finally, from aerial photography to Landsat to AirSAR, these and other remote-sensing technologies are often best used in concert, either merged, layered or draped together within an image-processing or GIS package. Or they can each serve as a preliminary check upon the other before the absolutely vital step of field verification (see Corbley 1999).

References

Adams, R. E. W. (1982). Ancient Maya canals: Grids and lattices in the Maya Jungle. *Archaeology, 35*(6), 28–35.

Blom, R., Cabrera, J., Clark, D., Comer, D., Godt, J., Golden, C., Hixson, D., Inomata, T., Irwin, D., Losos, E., Murtha, T., Pope, K., Quilter, J., Ringle, B., Saatchi, S., Scatena, F., Sever, T., & Sharer,

R. (2003). *NASA-AIRSAR campaign in Central America for archeological and conservation applications: A data acquisition research proposal.* Document prepared following the AirSAR Mesoamerica Committee Meeting held at Dumbarton Oaks in Washington, DC on 22 Oct 2003.

Corbley, K. P. (1999). Pioneering search for a primitive city. *GeoInfo Systems, 9*(6), 30–34.

Folan, W., Marcus, J., & Miller, W. F. (1995). Verification of a Maya settlement model through remote sensing. *Cambridge Archaeological Journal, 5*(2), 277–301.

Garrison, T. G., Chapman, B., Houston, S., Román, E., & Garrido López, J. L. (2011). Discovering ancient Maya settlements using airborne radar elevation data. *Journal of Archaeological Sciences, 38*(7), 1655–1662.

Garrison, T. G., Houston, S. D., Golden, C., Inomata, T., Nelson, Z., & Munson, J. (2008). Evaluating the use of IKONOS satellite imagery in lowland Maya settlement archaeology. *Journal of Archeological Science, 35*(10), 2770–2777.

Garza Tarazona de Gonzalez, S., & Kurjack, E. B. (1980). *Atlas Arqueologico del Estado de Yucatan* (Vol. 2). Mexico: Instituto Nacional de Antropologia e Historia, Centro Regional del Sureste.

Hixson, D. R. (2005). Measuring a Maya metropolis: The use of remote sensing for settlement pattern research at the classic Maya site of Chunchucmil, Yucatan, Mexico. *Institute of Maya Studies Newsletter, 34*(1), 1–4.

Hixson, D. R. (2011). *Settlement patterns and communication routes of the Western Maya Wetlands: An archaeological and remote-sensing survey, Chunchucmil, Yucatan, Mexico.* Unpublished Ph. D. Dissertation. Tulane University, New Orleans.

Hutson, S., Hixson, D., Dahlin, B. H., Magnoni, A., & Mazeau, D. (2008). Site and community at Chunchucmil and ancient Maya urban centers. *Journal of Field Archaeology, 33*(1), 19–40.

Lillesand, T. M., & Kiefer, R. W. (2000). *Remote sensing and image interpretation* (4th ed.). New York: Wiley.

Pope, K. O., & Dahlin, B. (1989). Ancient Maya wetland agriculture: New insights from ecological and remote sensing research. *Journal of Field Archaeology, 16*(1), 87–106.

Pope, K. O., Rejmankova, E., & Paris, J. F. (2001). Spaceborne imaging radar-C (SIR-C) observations of groundwater discharge and wetlands associated with the Chicxulub impact crater, northwestern Yucatan Peninsula, Mexico. *GSA Bulletin, 13*(3), 403–416.

Pope, K. O., Rey-Benayas, J. M., & Paris, J. F. (1994). Radar remote sensing of forest and wetland ecosystems in the Central American tropics. *Remote Sensing of Environment, 48*, 205–219.

Saturno, W., Sever, T., Irwin, D., Howell, B., & Garrison, T. (2007). Putting us on the map: Remote sensing investigation of the ancient Maya landscape. In J. Wiseman & F. El-Baz (Eds.), *Remote sensing in archaeology: Interdisciplinary contributions to archaeology* (pp. 137–160). New York: Springer Science and Business Media.

Sever, T. L. (1998). Validating prehistoric and current phenomena upon the landscape of the peten, Guatemala. In D. Liverman, E. F. Moran, R. R. Rindfuss, & P. C. Stern (Eds.), *People and pixels: Linking remote sensing and social science* (pp. 145–163). Washington, DC: National Academy Press.

Vlcek, D. T. (1978). Muros de delimitacion residencial en chunchucmil. *Boletin de la Escuela de Ciencias, Antropologicas de la Universidad de Yucatan, 28*, 55–64.

Vlcek, D. T., Garza de Gonzalez, S., & Kurjack, E. B. (1978). Contemporary farming and ancient Maya settlements: Some disconcerting evidence. In P. D. Harrison & B. L. Turner (Eds.), *Pre-Hispanic Maya agriculture* (pp. 211–223). Albuquerque: University of New Mexico.

Chapter 12
The Promise and Problem of Modeling Viewsheds in the Western Maya Lowlands

Charles Golden and Bryce Davenport

Abstract This paper addresses the benefits and challenges of modeling viewsheds with ASTER, SRTM, and AirSAR DEMs as they relate to human scale behaviors through a case study that examines the cultural and political significance of viewsheds for Precolumbian Maya rulers. The goal of this chapter is twofold: to illustrate the dramatic differences in currently available datasets for calculating viewsheds and to reflect on the implications indigenous concepts of vision and intervisibility hold for the reconstruction of ancient vistas. We conclude that movement through and vistas across the landscape participated in the construction of political power and authority in Classic period (c. 250–900 C.E.) Maya kingdoms, but that to achieve a reasonable quantitative model of these vistas the parameters of datasets available to most archaeologists are insufficient. Coarse-resolution DEMs used to model viewsheds vastly overstate the perceptible area, yet may conversely obscure areas that would otherwise be visible to a human observer. Visibility analyses have become a part of archaeological standard practice without due consideration of the cultural context of perception, or the resolution of data approximating human scales.

Keywords Viewsheds • Landscapes • Maya • AirSAR • ASTER • SRTM

12.1   Introduction

Viewshed and intervisibility studies have become popular applications of GIS software in many disciplines, including development planning and heritage conservation. Visuality analyses in archaeological research predate GIS platforms, and continue to carry a prodigious body of criticism about the theoretical and technical implications of modeling perception (reviewed in Wheatley and Gillings 2000). In this paper, we will briefly examine the cultural and political significance of viewsheds for Precolumbian Maya rulers, and then address the benefits and

D.C. Comer and M.J. Harrower, *Mapping Archaeological Landscapes from Space*,
SpringerBriefs in Archaeology, DOI 10.1007/978-1-4614-6074-9_12,
© Springer Science+Business Media New York 2013

challenges of modeling viewsheds with ASTER, SRTM, and AirSAR DEMs as they relate to human scale behaviors.

While early quantitative work modeled visibility as an attribute present in the environment (Fraser 1983), more recent approaches have sought to parse out the spatial possibility of intervisibility from perceptive acts by human agents (Tilley 1994; see Tschan et al. 2000). Following the humanistic geography of scholars such as Dennis Cosgrove (1984), viewing is a human engagement with the landscape. These computer modeling techniques, on the other hand, are potentially overly-deterministic abstractions that derive more from processing power than human experience. Although often noted as a hazard of spatial representation and analysis in post-structural theory, solutions to the problems of synopticism and 'objective viewers' have not seen much substantial methodological work (cf. Llobera 1996, 2003; Riggs and Dean 2007). Our goal in this chapter is twofold: to illustrate the dramatic differences in currently available datasets used for calculating viewsheds, and to reflect on the implications of indigenous concepts of vision and intervisibility for the reconstruction of ancient vistas.

12.2 Background

From the fourth through ninth centuries C.E., the rulers of the Maya kingdom centered on the archaeological site of Yaxchilan, Mexico vied with neighboring dynasties for political domination of the middle Usumacinta River valley (Figs. 12.1 and 12.2). In this competition the physiography of the land – a rugged karst terrain of steep hills, cliffs, sink holes, swamps and lakes – was a critical component of political action and not a passive canvas for the expansion of dynastic authority. The vistas offered by the landscapes surrounding the political capitals shaped notions of political space that extended across the polity. To reconstruct these vistas and the political role that they played, we must rely on remotely sensed data that ostensibly permit the modeling of human-scale viewsheds across the landscape.

Using epigraphic data and least-cost routes based on the 30 m (ASTER) and 90 m (SRTM) DEMs, the political and economic significance of the landscape seems immediately clear in terms of overland movements through the restricted valleys of the region. Routes of access into the core of the Yaxchilan kingdom from the Central Petén, the political heartland of the first-millennium Southern Maya Lowlands, as well as to marine resources from the north and south and volcanic materials such as obsidian and basalt from the highlands is limited to a few overland and challenging riverine corridors. These corridors were contested through marriage alliances and warfare by the rulers of Yaxchilan and its neighboring kingdoms throughout the Classic period.

Movement across this broken landscape is reasonably well-modeled at the resolution of the ASTER and SRTM data, and modern foot and automobile traffic follow very many of the least-cost routes modeled using these data. However, vistas and viewsheds in such terrain are also of profound cultural and political import.

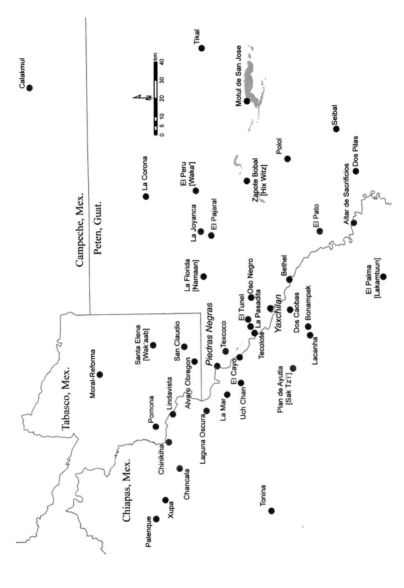

Fig. 12.1 Map of the region discussed showing the location of Yaxchilan

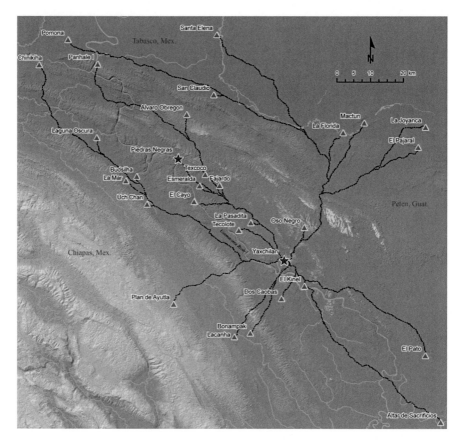

Fig. 12.2 Map of the Yaxchilan region with viewshed modeled using Airsar DEM (*yellow*) and least cost routes from neighboring sites modeled using ASTER DEM (*black dashed lines*)

These are more reasonably modeled with elevation models whose finer resolution approaches human-scale perception.

12.3   Understanding Maya Viewsheds: *Y-ichnal*

Vistas and viewsheds, as part and parcel of landscapes, are produced through social action and are profoundly political (Cosgrove 1984; Ingold 1993; Lefebvre 1991; Smith 2003). Ideology is embodied, contested and restructured in the organization of places: buildings, fields, forests, and workshops. The physical facts of the terrain – topography, hydrology, and vegetation – provide constraints, but by no means truly limit the creation of vistas. Vistas reveal to the viewer a framed, encapsulated and inherently circumscribed image of political organization.

To understand the particular political significance of vistas in the Maya Lowlands, we need to understand that for the Precolumbian Maya, sight was not passively received light and imagery, but was instead active and went out from the eye (Houston, Stuart, and Taube 2006: 167). Its political significance is emphasized epigraphically in the phrase *y-ichnal*. Houston, Stuart, and Taube (2006: 173–175) define y-ichnal as a creative visual field, rhetorically possessed in the inscriptions of the Classic period *only* by deities or by super-ordinate rulers: It was his/her/its -ichnal (Hanks 1993: 149; Herring 2005: 54–56; Houston and Taube 2000). Subordinates acceded in the ichnal of their overlords, overlords acceded in the ichnal of patron deities. Further, it was a dynamic and participatory sensory field that involved a reciprocal relationship between the ruler who possessed the -ichnal and all of the individuals within the -ichnal. Indeed, these other individuals could expand and extend the -ichnal of the ruler or deities (Houston and Taube 2000: 288; Houston, Suart, and Taube 2006: 174). The architectural settings of palaces, plazas, and temples, in turn, helped to shape the -ichnal by restricting and guiding vision and other sensory inputs.

Inscriptions also indicate that the depictions of individuals were indivisible from the personhood of the individuals themselves (Houston and Stuart 1998; Stuart 1996). Kings, queens, nobles and deities were perpetually present, perpetually in performance, and perpetually *viewing*. Thus, the creative, powerful visual field of the ruler depicted on the monument extended not merely from the biological person of the ruler, but also from every monument on which that ruler is depicted. The maintenance and extension of the ruler's visual field constituted a critical component of the maintenance and extension of rulership. Breaks in this visual field offered opportunities for disruption in, and contestation of, the authority of dynastic rulers and this had a profound effect on architectural arrangements within sites, site distributions across the landscape, and depictions of rulers and subordinate nobles on monuments (Golden 2010).

12.4 SRTM, ASTER, AirSAR: Approaching a Human Perspective

At their most basic, visibility analyses are binary queries, where a given raster cell is classified as in-view or out-of-view from the starting location based on their relative elevations. The total set of cells visible from the initial point of reference is described as its viewshed. GIS packages frequently give options for viewer and target offset (height above surface), maximal and minimal distances, and limitations on horizontal and vertical viewing angles, and many authors have suggested further computations to better simulate the variability of viewer height and acuity, climatic and temporal conditions, topographic orientation and directional preferences, and approximations based on the granularity of the source material (e.g. Fisher 1996a; Higuchi 1983).

The pertinent values for viewshed analyses are typically derived from a digital elevation model (DEM) raster image. Wheatley and Gillings (2000: 9) note that as visibility studies are analyses of sample-based models, the precision, accuracy, and resolution of the DEM are of primary concern in any study of human-scale phenomena (Doyle et al. 2012). While in most cases archaeologists are not themselves responsible for the data used in DEM production, the wide availability of these products and the relative ease of their manipulation in modern GIS packages has enabled a plethora of multiscalar research programs without accompanying theoretical concern for the level of spatial representation (Lock and Harris 2000: xviii–xxi).

In order to illustrate the dramatic effect that changes in scale have on the extent of visible areas, the viewsheds were modeled from equivalent viewpoints (centered on the South Acropolis at Yaxchilan, the site's highest architectural point) in DEMs developed from three remote sensing platforms, ASTER, SRTM, and AIRSAR (Figs. 12.3, 12.4, and 12.5). Advanced Spaceborne Thermal Emission and Reflection Radiometer (ASTER) products are hyperspectral (15 bands) images from three sensors – visible and near infrared, at 15 m resolution; short wave infrared, at 30 m; and thermal infrared, at 90 m. ASTER DEMS are derived from the near infrared bands and downsampled to 30 m resolution. While the spectral coverage afforded by these instruments means that there is great potential for ASTER analyses of soil and vegetation composition as proxies for archaeological remains (Parcak 2009: 69), canopy readings can complicate the production of elevation models in densely vegetated areas like the Usumacinta River Valley. The Shuttle Radar Topography Mission was originally conceived with the production of digital elevation and terrain models in mind. The mission created a 90 m resolution global DEM using C-band radar, although this range also has weak penetrative qualities in dense vegetation. SRTM data are well suited to larger-scale terrain reconstructions such as paleo-hydrology and cost-surface analyses. Airborne Synthetic Aperture Radar is an aircraft-based platform that also uses C-band radar waves to record topography, with detail from L- and P-bands for surface structure; while this combination allows for greater terrain documentation beneath dense vegetation, the low altitude and side-looking implementation of AIRSAR flights can create issues of distortion and "data shadows" in areas of high relief (Potsis et al. 2000; Sano et al. 2005). Although the original AIRSAR platform delivered resolutions of 12–15 m, more recent missions, such as the 2004 AirSAR Mesoamerica Campaign, delivered 3–5 m resolution images and 5 m resolution DEMs (see Garrison et al. 2011 and Hixson, this volume, for application in the Maya area).

The horizontal resolution of the raster image translates into the distance averaged into the values of a single pixel. When considering the values used to produce elevation models, areas of moderate to very rugged terrain are unlikely to maintain consistent elevations over the resolution provided by freely available data such as SRTM and ASTER DEMs. Equally, the minimum resolutions of these products are much larger than the perspective of a human observer, complicating both the placement of the viewpoint and the viewable area. To demonstrate the effect that these different resolutions have on visibility, three binary viewsheds were run with the acropolis of Yaxchilan as the viewing point. While simple yes-or-no queries are not best practice in visibility analyses (Fisher 1996a), the simplicity of presentation more clearly demonstrates the changes to the hypothetical viewable area at each resolution.

The SRTM-derived DEM, at 90 m resolution, produced a viewshed of 136,412 sq km (Fig. 12.3). Consistent with the rugged topography of the region, visibility is

Fig. 12.3 Viewshed (*yellow*) from the South Acropolis at Yaxchilan, modeled using the SRTM DEM (base map image is ASTER DEM, not used in modeling viewshed)

fragmented and restricted to mountain slopes and closer valley basins. The 30 m ASTER-derived DEM produced a viewshed of 66,741 sq km (Fig. 12.4), or just under half of the visible area in the SRTM viewshed. While the reduction in scope is substantial, another interesting point to note is that a formerly obscured area to the southwest of Yaxchilan's acropolis is now visible. Finally, the 5 m resolution AIRSAR derived DEM returned 36,975 sq km of visible area (Fig. 12.5), just over half of the ASTER viewshed and just over a quarter of that of the SRTM. While the vista to the north and east is substantially reduced, the area to the southwest has increased.

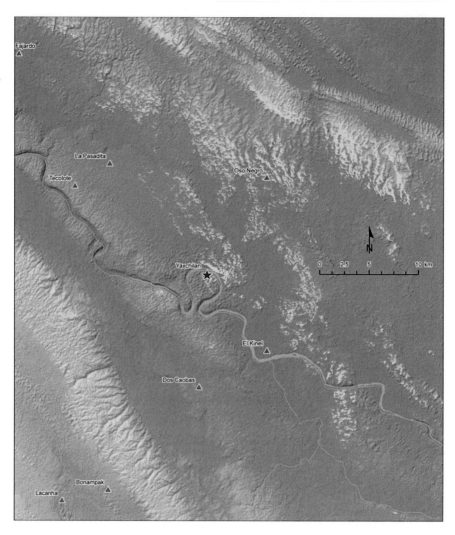

Fig. 12.4 Viewshed (*yellow*) from the South Acropolis at Yaxchilan, modeled using the ASTER DEM (base map image is ASTER DEM)

12.5 Approaching Ancient Viewsheds

What, then, does a more human-scale model like that produced using AirSAR data suggest? The layout of the dynastic center at Yaxchilan is outward looking (Tate 1992). It opens to the river and to the wide vistas offered eastward across the river toward the Sierra del Lacandón. Travelers turned a bend in the river and, with trees removed from the picture, would immediately be able to take in a waterfront of monumental structures. A broad open plaza strewn with buildings dominates the lower

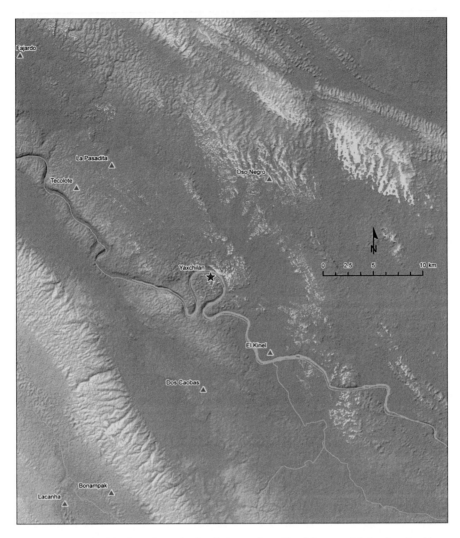

Fig. 12.5 Viewshed (*yellow*) from the South Acropolis at Yaxchilan, modeled using the Airsar DEM (base map image is ASTER DEM, not used in modeling viewshed)

sections of the epicenter. At high water mark, the traveler is at eye level with much of the dynastic center, unavoidably encompassed in the view of its residents and the ichnal of its lords and deities. The vistas, particularly from the South Acropolis across the river, would have opened up through gaps in the low hills to encompass within the-ichnal of Yaxchilan's rulers the valley floor running all the way to the Sierra del Lacandón, some 13 km away, and a gap leading out to the sometime enemy/ sometime allied kingdom of Hix Witz centered on the archaeological site now known as Zapote Bobal (Breuil-Martinez et al. 2004; Gámez et al. 2007). Looking westward from the flanks of the sierra, Yaxchilan is quite visible on a clear day.

Moreover, many of Yaxchilan's subordinate political centers are intervisible with the capital. Hilltop redoubts that loom above a system of defensive walls at the northern border center of Tecolote would have allowed for signaling between Tecolote and Yaxchilan. Also visible from Yaxchilan was the architectural epicenter of La Pasadita (Fig. 12.6), where the principal structure originally housed three lintels depicting the ruler of Yaxchilan together with governor of La Pasadita conducting ceremonies, receiving tribute, and dominating a captive prince from neighboring Piedras Negras (Martin and Grube 2008: 131; Nahm 1997). Other secondary centers, including Oso Negro which guards the pass opening eastward towards the Hix Witz kingdom and the Central Petén, were visible from the capital itself, and were intervisible with one another.

Further, the rulers of Yaxchilan were cosmopolitan, making appearances on monuments at secondary centers throughout the kingdom, thus expanding their vision, their-ichnal and their power, perpetually into the countryside. La Pasadita is particularly interesting. Although its epicenter is less architecturally imposing than that of neighboring Tecolote, it is the source of at least four sculpted lintels while no such monuments have yet been associated with Tecolote. We suggest that the presence of Yaxchilan's ruler on monuments at La Pasadita's principal palace structure is directly related to this building's intervisibility with Yaxchilan's central architecture which effectively extended the king's visual field northward, out into that border zone (Fig. 12.6). The palace structures at Tecolote, in contrast, are not intervisible with Yaxchilan. Instead, the hilltop redoubts that form part of the complex of border defense features at Tecolote are intervisible with La Pasadita, Yaxchilan, and other borderland sites (Scherer and Golden 2009).

12.6   Conclusion

Given myriad possibilities for engaging with the landscape, the dynastic rulers of Yaxchilan turned their vision outwards. Vistas across the landscape did not closely prescribe architectural form, site placement and monument composition in these two polities. Rather, movement through and vistas across the landscape participated in the construction of political power and authority in these Maya kingdoms.

To achieve a reasonable modern model of these vistas, however, the parameters of datasets available to most archaeologists are insufficient. On the one hand, using coarse-resolution DEMs to model viewsheds vastly overstates the perceptible area over the landscape, yet can also obscure areas that would otherwise be visible. We caution that the wide availability of these products and the routine inclusion of viewshed functions in commercial and open source GIS platforms run the risk of making visibility analyses part and parcel of archaeological standard practice without due consideration of the cultural context of perception, or the resolution of the data approximating human scales. The concordance between visible sites, inscriptions, and monuments in the Yaxchilan kingdom is a primary feature of the northern border

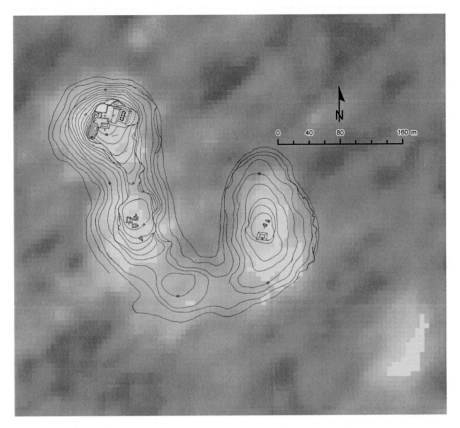

Fig. 12.6 Architectural epicenter of La Pasadita (as mapped by Ian Graham) over Airsar DEM. *Yellow* indicates pixels visible from the South Acropolis at Yaxchilan when modeled using Airsar DEM. The largest building at the north end of the site once contained three carved monuments, at least two of which depicted the ruler of Yaxchilan

only, not a hard and fast rule of settlement patterning across the entire kingdom. This emphasis on a panoptic view of the border was likely a component of the centuries-long contest between Yaxchilan and its northern neighbor, the kingdom of Piedras Negras, for control of the riverine and overland passages through the Middle Usumacinta River Basin (Anaya Hernández 2001, 2005; Anaya Hernández et al. 2003; Aliphat Fernandez 1994, 1996; Canter 2007; Golden and Scherer 2006; Golden et al. 2008; Scherer and Golden 2009). Modeling topography is a necessary but insufficient feature of reconstructing ancient landscapes, and the role the visibility plays, even in historically grounded case studies, varies at the interstices of environment and polity. Above and beyond the question of whether a site on the landscape is visible, we underscore the importance of understanding the logic and practice of political vistas as a tool of governance.

Acknowledgements Field research in Guatemala conducted by the Sierra del Lacandón Regional Archaeology Project directed by Charles Golden, Andrew Scherer, Rosaura Vasquez, Ana Lucia Arroyave, and Luz Midilia Marroquin with permissions from the Insituto de Antropología e Historia of Guatemala (IDAEH). Financial support provided by the Foundation for the Advancement of Mesoamerican Studies, Inc. (Grants #02020, 05027, and 07043), the National Geographic Society (Grants #7575-04, and 7636-04), the National Science Foundation (BCS-0406472, BCS-0715463, BCS-1115818), Dumbarton Oaks Research Library, H. John Heinz III Charitable Trust Grants for Latin American Archeology, the Kaplan Fund through the World Monuments Fund, Brown University, the Norman Fund and the Jane's Fund at Brandeis University, Wagner College, and Baylor University.

References

Aliphat Fernandez, M. M. (1994). *Classic Maya landscape in the Upper Usumacinta River Valley*. Ph. D. Dissertation, Department of Anthropology, University of Calgary.

Aliphat Fernandez, M. M. (1996). Arqueología y Paisajes del Alto Usumacinta. *Arqueología Mexicana, 4*(22), 24–29.

Anaya Hernández, A. (2001). *Site interaction and political geography in the upper Usumacinta region during the late classic: A GIS approach* (Bar international series, Vol. 994). Oxford: J. and E. Hedges.

Anaya Hernández, A. (2005). Strategic location and territorial integrity: The role of subsidiary sites in the classic Maya Kingdoms of the Upper Usumacinta Region. *Internet Archaeology 19*. http://intarch.ac.uk/journal/issue19/anaya_index.html. Accessed 11 June 2012

Anaya Hernández, A., Guenter, S. P., & Zender, M. U. (2003). SakTz'i', a classic Maya center: A locational model based on GIS and epigraphy. *Latin American Antiquity, 14*(2), 179–191.

Breuil-Martinez, V., Gamez, L., Fitzsimmons, J., Metailie, J. P., Barrios, E., & Roman, E. (2004). Primeras Noticias de Zapote Bobal, Una Ciudad Maya Clasica del Norocidente de Peten, Guatemala. *Mayab, 17*, 61–83.

Canter, R. L. (2007). Rivers among the ruins: The Usumacinta. *The PARI Journal, 7*(3), 1–24.

Cosgrove, D. (1984). *Social formation and symbolic landscape*. London: Croom Helm.

Doyle, J. A., Garrison, T. G., & Houston, S. D. (2012). Watchful realms: Integrating GIS analysis and political history in the southern Maya lowlands. *Antiquity, 86*, 792–807.

Fisher, P. (1996a). Extending the applicability of viewsheds in landscape planning. *Photogrammetric Engineering and Remote Sensing, 62*(11), 1297–1302.

Fraser, D. (1983). *Land and society in Neolithic Orkney*. Oxford: BAR 117.

Gámez, L., Fitzsimmons, J., & Forné, M. (2007). Epigrafía y Arqueología de Hixwitz: Investigaciones en Zapote Bobal, La Libertad, Petén. In J. P. Laporte, B. Arroyo, & H. Mejia (Eds.), *XX Simposio de Investigaciones Arqueológicas en Guatemala, 2006* (pp. 345–367). Guatemala: Museo Nacional de Arqueología y Etnología.

Garrison, T. G., Chapman, B., Houston, S., Roman, E., & Garrido Lopez, J. L. (2011). Discovering ancient Maya settlements using airborne radar elevation data. *Journal of Archaeological Science, 38*(7), 1655–1662.

Golden, C. (2010). Frayed at the edges: The re-creation of histories and memories on the frontiers of classic period Maya polities. *Ancient Mesoamerica, 21*(2), 373–384.

Golden, C., & Scherer, A. K. (2006). Border problems: Recent archaeological research along the Usumacinta river. *The PARI Journal, 7*(2), 1–16.

Golden, C., Scherer, A. K., René Muñoz, A., & Vásquez, R. (2008). Piedras negras and yaxchilan: Divergent political trajectories in adjacent Maya polities. *Latin American Antiquity, 19*(2), 249–274.

Hanks, W. F. (1993) Metalanguage and the pragmatics of deixis. In J. Lucy (Ed.) *Reflexive Language: Reported speech and metapragmatics* (pp. 127–158). New York: Cambridge University Press.

Herring, A. (2005). *Art and writing in the Maya cities, A.D. 600–800: A poetics of line*. New York: Cambridge University Press.

Higuchi, T. (1983). *Visual and spatial structure of landscapes*. Cambridge: Massachusetts Institute of Technology.

Houston, S. D., & Stuart, D. (1998). The ancient Maya self: Personhood and portraiture in the classic period. *Res, 33*, 73–101.

Houston, S.D., Stuart, D., & Taube, K.A. (2006). *The Memory of Bones: Body, Being, and Experience among the Classic Maya*. Austin: University of Texas Press.

Houston, S. D., & Taube, K. A. (2000). An archaeology of the senses: Perceptions and cultural expression in ancient Mesoamerica. *Cambridge Archaeological Journal, 10*, 261–294.

Ingold, T. (1993). The temporality of the landscape. *World Archaeology, 25*, 152–174.

Lefebvre, H. (1991). *The production of space*. Oxford: Basil Blackwell.

Llobera, M. (1996). Exploring the topography of mind: GIS, social space and archaeology. *Antiquity, 70*, 612–622.

Llobera, M. (2003). Extending GIS-based visual analysis: The concept of *viewscapes*. *International Journal of Geographical Information Science, 17*(1), 25–48.

Lock, G., & Harris, T. (2000). Introduction: Return to ravello. In G. Lock (Ed.), *Beyond the map: Archaeology and spatial technologies* (pp. xiii–xxv). Oxford: Ios Press.

Martin, S., & Grube, N. (2008). *Chronicle of the Maya kings and queens*. New York: Thames and Hudson.

Nahm, W. (1997). Hieroglyphic stairway 1 at yaxchilan. *Mexicon, 19*, 65–69.

Parcak, S. (2009). *Satellite remote sensing for archaeology*. New York: Taylor and Francis.

Potsis, A., Uzunoglou, N., Frangos, P., Horn, R., & Lumprecht, K. (2000). Analysis of P-band synthetic sperture radar for airborne and spaceborne applications. In *Paper presented at the RTO SET symposium on space-based observation technology on the island of Samos*, Greece, 16–18 Oct 2000, and published in RTO MP-61.

Riggs, P. D., & Dean, D. J. (2007). An investigation into the causes of errors and inconsistencies in predicted viewsheds. *Transactions in GIS, 11*, 175–196.

Sano, E., Ferreira, L., & Huete, A. (2005). Synthetic aperture radar (L band) and optical vegetation indices for discriminating the Brazilian savanna physiognomies: A comparative analysis. *Earth Interactions, 9*, 15.

Scherer, A. K., & Golden, C. (2009). Tecolote, guatemala: Archaeological evidence for a fortified late classic Maya political border. *Journal of Field Archaeology, 34*(3), 285–304.

Smith, A. T. (2003). *The political landscape: Constellations of authority in early complex polities*. Berkeley: University of California Press.

Stuart, D. (1996). Kings of stone: A consideration of ancient stelae in Maya ritual and representation. *Res, 29–30*, 148–171.

Tate, C. E. (1992). Yaxchilan: The Design of a Maya Ceremonial City. Austin: University of Texas Press.

Tilley, C. (1994). *A phenomenology of landscape*. Oxford: Berg.

Tschan, A., Raczkowski, W., & Latasowa, M. (2000). Perception and viewsheds: Are they mutually inclusive? In G. Lock (Ed.), *Beyond the map: Archaeology and spatial technologies* (pp. 29–48). Oxford: Ios Press.

Wheatley, D., & Gillings, M. (2000). Vision, perception and GIS: Developing enriched approaches to the study of archaeological visibility. In G. Lock (Ed.), *Beyond the map: Archaeology and spatial technologies* (pp. 1–27). Oxford: Ios Press.

Chapter 13
The Influence of Viewshed on Prehistoric Archaeological Site Patterning at San Clemente Island as Suggested by Analysis of Synthetic Aperture Radar Images

Douglas C. Comer, Ronald G. Blom, and William Megarry

Abstract The use of synthetic aperture radar (SAR) imagery collected by the NASA AirSAR platform to detect archaeological site location on San Clemente Island, one of the Southern Channel Islands, has been described in several publications. Here we provide a concise review of the use of SAR for that purpose, as well as a description of how a precise surface model also generated by SAR data was used to examine the spatial distribution of sites by means of viewshed modeling. Results of the direct detection model are evaluated using gain statistics that make use of a recently completed 100% survey of all accessible areas on San Clemente Island. Viewshed analyses generated by use of the surface model suggest a number of economic and ideological factors that might have influenced the distribution of archaeological sites on the island. Among these are intervisibility among locations on both San Clemente Island and nearby Santa Catalina Island that provided the means by which to coordinate crucial substance activities, such as hunting sea mammals, and reinforced the social relationships that were essential to such activities.

Keywords Synthetic aperture radar (SAR) • Detection of archaeological sites • Viewshed analysis • California Channel Islands • Aerial and satellite remote sensing in archaeology

13.1 Introduction

The use of synthetic aperture radar (SAR) imagery collected by the NASA AirSAR and GeoSAR platforms to detect archaeological sites on San Clemente Island, one of the California Southern Channel Islands (see Fig. 13.1), has been described in several publications (Comer and Blom 2007a, b, c; Comer 2008). The research made use of an exceptionally rich set of SAR imagery, generated from multi-band and multi-polar data. Since reporting our results in 2007 and 2008, the Navy has surveyed all accessible portions of the island. Using the results of this survey, we present here gain statistics (Kvamme 1988) strongly indicating that the site detection

D.C. Comer and M.J. Harrower, *Mapping Archaeological Landscapes from Space*, SpringerBriefs in Archaeology, DOI 10.1007/978-1-4614-6074-9_13, © Springer Science+Business Media New York 2013

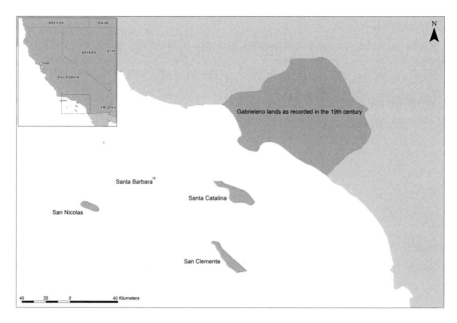

Fig. 13.1 Location map of San Clemente Island, other Southern Channel Islands, and the area historically occupied by the Gabrielino on the mainland

protocols we devised to analyze SAR images were extremely productive. In addition, the distribution of archaeological sites as seen in the probabilistic detection model suggests a number of environmental and cultural influences that might have influenced site selection. This distribution is examined through the use of another SAR product, a high-resolution surface model generated by interferometric analysis of a SAR band.

13.2 Site Distribution and Viewshed

Viewsheds from sites on San Clemente Island were calculated with a high-resolution surface model (approximately 3-m postings and elevation precision of about 1.5 m) that was generated by interferometric analysis of one of the SAR bands. In recent years, several archaeologists have argued that viewsheds from archaeological sites have played important roles in ancestor veneration, developing shared memories, strengthening social ties, delineation of territories, and structuring access to resources. Bongers et al. (2012) have examined viewsheds from *chalupas* (above-ground burial tombs) in the Lake Titicaca basin of Peru during the Late Intermediate Period. A statistical test established that cumulative *viewsheds from* chalupas was much greater than that from random locations and that *visibility of* the chalupas from settlements was different from viewsheds from random locales.The same held true for viewsheds from Lake Umayo, an especially rich resource-extraction area. Thus, there seemed to be a preference for settling on and exploiting the resources of places from which chalupas could be seen.

Fisher and Farrelly (1997) provide another example with their work on the island of North Mull in Scotland. Their tests suggested that certain views from cairns on that island were more important than others; among them, the views to the sea across to the Island of Ulva, the sea to the north, and the view down and across the Sound of Mull. They note that most of the water-borne trade occurred in these areas, and speculate that beyond economic importance, these views are an element in the social/ritual focus of the Bronze Age builders of the cairns. In his review of how archaeologists have either sidestepped or used religious ritual in the analysis of their findings, Fogelin (2007) notes that some regards the question "Is it religious?" is fundamentally flawed. He argues that ritual and religion act to inculcate modes of behavior and to transmit information from one generation to the next. In what follows, we suggest that viewsheds from clusters of sites on the San Clemente Island plateau might very well have functioned in this way. They are located at high elevations away from the abundant marine resources on the west and north coasts of the island.

13.3   San Clemente Island

The Southern Channel Islands of San Clemente Island, Santa Catalina Island, and San Nicolas Island, as well as the Palos Verdes Peninsula, including areas in southern and eastern Los Angeles and northern Orange Counties, have been occupied by populations that spoke a common Uto-Aztecan dialect since at least European contact (see Figure 13.1). Europeans referred to these populations as Gabrielino. Human occupation of the Southern Channel Islands dates to long before this; human remains on Santa Catalina Island have yielded dates of 12,000 B.P. Many radiocarbon dates of approximately 10,000 B.P. have been obtained from the Eel Point site on San Clemente Island. The Southern Channel Islands have never been connected to the mainland; therefore, seafaring among the islands and the mainland throughout the intervening millennia is strongly indicated (Goldberg et al. 2000: 31). The oldest of the types of habitation sites detected by our aerial and satellite remote sensing research are 4,000–5,000 B.P. and display the same patterning as more recent ones, suggesting that a similar set of factors influenced site selection over millennia.

San Clemente Island was probably never as densely populated as Santa Catalina Island because it receives just half the precipitation of the latter. Nevertheless, the site density at San Clemente Island is much higher than that of the other Channel Islands or on the mainland (Meighan 2000: 8). This, among other evidence, suggests that San Clemente Island has been inhabited continuously for as long as humans have been on the Southern Channel Islands (Goldberg et al. 2000: 40).

13.4   The Detection of Sites Using Synthetic Aperture Radar

Synthetic aperture radar (SAR) image data is obtained by transmitting and receiving different frequencies, or bands, of radar from a moving platform such as an airplane or satellite. Movement creates a synthetically large antenna that improves resolution in the along-track direction. Thus, a rapidly moving aircraft carrying SAR apparatus can survey wide areas quickly while detecting targets of interest more precisely than conventional radar.

SAR can differentiate among both objects and material characteristics. Topography, electrical permittivity (measurable responses of a material to electric fields), roughness, and geometry (structure) of the target affect radar echo. Because radar can penetrate many materials to depths of varying degrees, SAR can detect features not visible in optical imaging, including those covered by clouds, vegetation, soil, and other materials. Further, SAR instruments can operate in two modes. In *conventional mode,* a SAR can look to either the left or the right side of the airborne or spaceborne platform, and generate a 2D image consisting of radar backscattering cross section values for each pixel within the image. In *interferometric SAR (InSAR) mode*, where there are two receiving antennas, the radar can generate the height of each pixel in addition to the 2D backscattering image of the conventional mode.

The SAR data utilized in our research was obtained by use of two platforms designed by JPL/NASA. The first was AirSAR, mounted on a DC-8 aircraft. AirSAR had the capability to collect multi-band (P-, L-, and C-band) data that could be polarized in several different ways. The other was GeoSAR, designed and built by JPL/NASA for use by the private sector as a part of a privatization program. Together they provided an enormously rich set of data that could be used to characterize the landscape of San Clemente Island and the human and natural features there.

SAR can detect the location of archaeological sites by identifying anomalous scattering of radar bands. A schematic representation of a typical San Clemente Island habitation site is seen in Fig. 13.2. As seen there, radar bands of different lengths are scattered by objects of different sizes – typically when they encounter objects about one-fourth to one-third of band length. Dark, organic midden soils are found in the sites, creating an environment in which vegetation differs in type and vigor from that which surrounds it. They also contain a scatter of fist-sized rocks. The P-band, about 75 cm in wave length, is unaffected by most of this material. The rich, moist, electrically conductive soils that developed during human occupation of the site, however, reflect P-band, much as a beam of light would be if shone on a mirror at an angle. The return to the sensor was therefore weak in comparison to backscatter strength from soils surrounding the sites. As seen in Fig. 13.2, the shorter SAR bands were scattered by rocks and vegetation at the site, but in different ways that produced varying magnitudes of return depending upon band length. Band polarization also affected strength of backscatter. Each band was transmitted both vertically and horizontally and received both vertically and horizontally. Polarized bands were scattered differently by different vegetative structure. Grasses, which have an essentially linear structure, generated strong returns from short bands

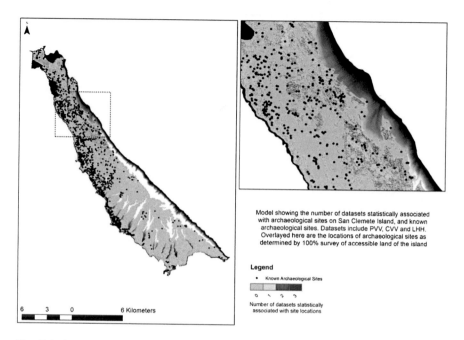

Fig. 13.2 Site patterning indicated by direct detection protocols used to analyze SAR datasets

polarized vertically at both transmission and reception, but very weak returns when transmitted vertically and received horizontally, and vice-versa. Each polarized band was scattered in different ways by the various characteristics of archaeological sites. Comparisons of backscatter strength were made between site and non-site areas for each polarized band. The degree of difference was great enough for several polarized bands to be statistically different to a 95% degree of certainty. The statistical protocol utilized to establish associations, or the lack thereof, between band return magnitude (datasets) and sites vs. non-sites was the Student's t-test (for a more comprehensive description of statistical protocols employed, see Comer and Blom, 2007a, and Tilton and Comer, this volume).

The distribution of sites as detected in this way have recently been compared with the locations of archaeological sites on San Clemente Island as recorded by a 100% inventory of all accessible areas on the island completed in the years since the 2007 publication of findings by Comer and Blom. As seen in Table 13.1, for more than 75% of San Clemente Island, returns from none of the SAR bands was statistically associated with the presence of archaeological sites. Accordingly, the gain statistic, calculated as 1 - % area / % sites (Kvamme, 1988), was therefore a large negative number for most of the island, indicating strongly that where sites were not detected, there were extremely few sites. Where two or more bands detected sites, the findings of the 100% survey were dramatically different. The gain statistic increased as more bands indicated the presence of archaeological sites. Two bands were associated with sites in an areas that comprised 3.33% of the island, and here

Table 13.1 Gain statistics for direct detection results by number of associated SAR bands

	Number of associated SAR bands	% of area	% of sites	Gain statistic
Statistical associations were	0	75.55375	15.6	−3.85252
with SAR returns as	1	20.92681	37.4	0.440908
recorded in 5-m pixels.	2	3.329719	38.7	0.956286
Sites on San Clemente	3	0.189723	8.3	0.997753
Island average approxi-				
mately 10 m in diameter				
and associated environmen-				
tal changes extend several				
more meters, therefore a				
20-m buffer provided the				
area associated with a site				
to a 95% confidence level.				
This was equivalent to				
re-sampling to 20 m pixels,				
not done because re-sam-				
pling alters return values				

almost 39% of all sites were found, producing a strikingly large gain statistic of almost 0.96. The maximum gain statistic, of course, would be 1.00, and where three bands were associated with the presence of sites, the gain statistic was practically that, or 0.998. In interpreting the importance of these scores, it should be noted that "traditional" methods for predicting the presence of archaeological sites, done by establishing historical associations between environmental factors and site locations, have seldom, if ever, produced comparable gain statistics. At least one expert has speculated that such models could never produce a gain statistic of more than 0.70 (Ebert 2000: 133). Future research in direct detection protocols will probably develop the means by which to identify the locations of a greater percentage of sites with similar degrees of certainly. In this case, however, we raise the additional interesting possibility that environmental changes produced by frequent human use of the island plateau might have been pronounced enough to generate false positives.

13.5   Site Distribution and Viewshed

13.5.1   Marine Viewshed

A relatively high density of sites was indicated by the SAR analysis on the northern plateau of the island (see bounded box in Fig. 13.3). These are some distance away from the most dependable sources of water today, which are in the ravines in the southern half of the island. Further, access to marine resources is not easy from this area of the plateau. Dense alignments of sites run along the western and northern

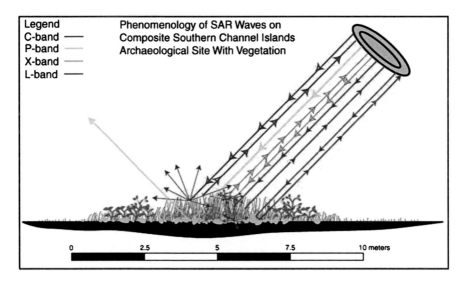

Fig. 13.3 Schematic of typical San Clemente Island archaeological site and SAR band interaction

coastlines, where there are many coves and beaches. These are seen also along the western edges of the marine terraces, not far from the coastline, with excellent views of the coastline and the ocean beyond. In contrast, access to the ocean on the east side of the island is extremely difficult if not impossible because of the escarpment there, a sheer cliff. The plateau does, however, provides a much greater accumulated view of the ocean than do sites located near ravines or with much easier access to the ocean (Fig. 13.4).

The diet of prehistoric San Clemente Island inhabitants was taken almost completely from the ocean, as evidenced by analyses of human bone collagen (Goldberg 1993) and nitrogen isotope analysis (Masters et al. n.d.). Porcasi et al. 2000 analyzed faunal remains taken from strata at the Eel Point site dating from 7,040 B.C. to 1,400 A.D. and found that only 1% were from terrestrial animal species. Of the 2,403 bones that could be identified, 38.4% were delphinidae (nearly all dolphins), 32.9% were pinnipeds, 20.8 were sea otters, and 7.9% were large cetaceans. An additional 7,715 bones could not be identified, but in light of the fact that only 1% of identified bones were terrestrial, virtually all unidentified bones were considered to be marine mammal. Clustering of sites at locations with optimal viewsheds of the waters surrounding the island is what one might expect given the manner in which sea mammals have been harvested in other regions of the world. Harvesting these mammals typically involves rapid and coordinated mobilization of people. When sea mammals congregate, island inhabitants must quickly find their way to sea craft, surround, and drive mammals into coves, where killed or wounded animals are thrown onshore for butchering.

Such hunts are known ethnographically, and take place today in, among other places, the Faroe Islands (Arms 2010: 102–106), the New Hebrides and, as recorded

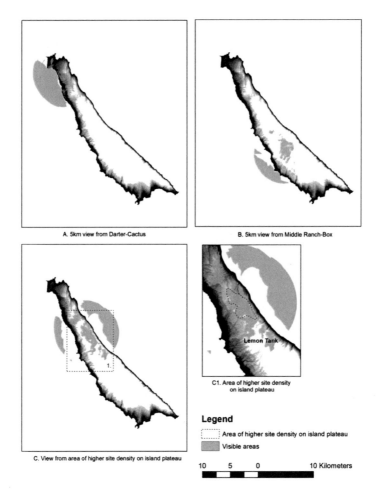

A. 5km view from Darter-Cactus

B. 5km view from Middle Ranch-Box

C. View from area of higher site density on island plateau

C1. Area of higher site density on island plateau

Lemon Tank

Legend

Area of higher site density on island plateau

Visible areas

10   5   0   10 Kilometers

Fig. 13.4 Viewsheds (extending only 5 km) from several San Clemente Island habitation site locales

in the Academy Award winning documentary "The Cove," in Taiji, Wakayama, in Japan. The earliest record of the Faroe Island hunt dates to 1584, and records show that these have taken place every year since 1709. By law, the hunters can use only the simplest of weapons and tools. Coordination is by means of bonfires visible on neighboring islands.

13.5.2   Viewsheds to and From Santa Catalina Island

The sites on the plateau seen in Fig. 13.3 are in an area with clear views to Santa Catalina Island (Fig. 13.5). The Spanish priest Geronimo Boscana wrote in the early nineteenth century that the largest settlement on Santa Catalina Island was at Two Harbors. Two Harbors therefore would also probably have been the location of the most important ritual site on Santa Catalina Island during the historic period. Boscana (1978: 15) wrote that temples erected on Santa Catalina in honor of the god Chinigchinich were always in the center of towns and contiguous to dwelling of the captain, or chief.

San Clemente Island, in contrast, had no towns, only scattered habitation sites. It is reasonable to assume that to the more scattered population on San Clemente Island, a political and religious center like Two Harbors would be of great interest. Intervisibility would have provided an experiential link between the populations on the two islands. Intervisibility would also have made communication possible. Communication would serve well when one of the two island groups observed dangers or opportunities, to coordinate activities, and to provide the means for joint participation in ceremonies and observances.

Chinigchinich sites have been identified and excavated on San Clemente Island. These are located within an area called Lemon Tank, just to the south of the sites on the plateau. The setting here is dramatic: It is adjacent to the eastern escarpment of the island, which plunges approximately 400 m to the ocean below. As moist air carried by winds from the west moves from sea level upslope to this location, thin clouds often form, scudding over the Chinigchinich sites, then plunging over the eastern escarpment like a waterfall. Figure 13.5 also illustrates that there is intervisibility among a few places on San Clemente Island and the Lemon Tank ritual sites, and two of these areas fall within the area of high site density on the plateau seen in Fig. 13.3 on San Clemente Island.

13.6 Location of Sites Within a Cost Surface

Figure 13.6 presents a cost surface for San Clemente Island that is calculated using the site called Eel Point as a starting point. Eel Point can be taken as representative of a coastal location that provides access to a great variety of marine resources. Coves offer habitat for shellfish and many fish species, kelp beds just off shore habitat for other fish species and sea otters, and the ocean floor descends dramatically a few tens of meters off the west coast of San Clemente Island. In these deep waters are larger species of fish as well as sea mammals. The coves themselves are places where sea mammals can be harvested, in the manner described above. The energetic cost of moving from Eel Cove to the densely settled portion of the island seen in Fig. 13.3 is moderate, as indicated by Fig. 13.6, and the cost of moving to the ritual sites at Lemon Tanks only slightly more. Costs increase rapidly as one moves upslope (to the southwest) from there. As seen in Fig. 13.7, viewsheds from these higher elevations do not seem to reward this additional cost. What can be seen from both the plateau sites and the highest point on the island is seen in pink. The area seen only from the highest point of the island is green; that which can be seen from plateau sites is tan, and areas not visible from either in white. Accumulated viewshed is

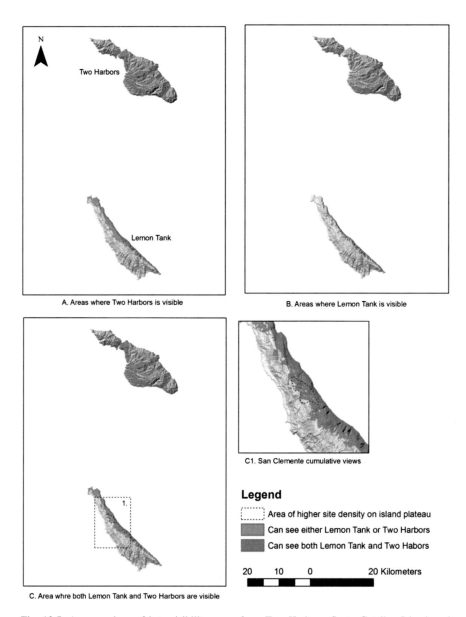

A. Areas where Two Harbors is visible

B. Areas where Lemon Tank is visible

C1. San Clemente cumulative views

Legend

⬚ Area of higher site density on island plateau

▨ Can see either Lemon Tank or Two Harbors

▨ Can see both Lemon Tank and Two Habors

20   10   0   20 Kilometers

C. Area whre both Lemon Tank and Two Harbors are visible

Fig. 13.5 A comparison of intervisibility areas from Two Harbors, Santa Catalina Island, and Lemon Tank, San Clemente Island

similar, but what can be better seen from the sites on the plateau was the rich and accessible west coast of the island. Areas that can be better seen from the highest point are essentially inaccessible because they are below the eastern escarpment.

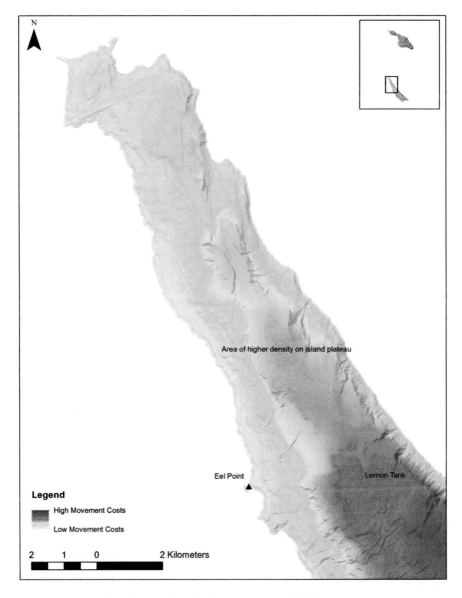

Fig. 13.6 Cost surface calculated from Eel Point archaeological site

13.7   Conclusion

Site detection is both an end in itself and a first step in the analysis of the archaeological record. Sites found can be excavated, of course, but the patterning of sites can also be informative about the environmental, social, and cultural forces that structure human occupation of a landscape. Site patterning at San Clemente

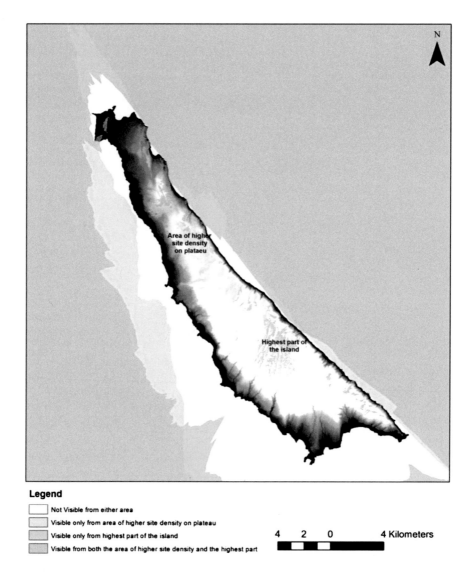

Legend

☐ Not Visible from either area

☐ Visible only from area of higher site density on plateau

☐ Visible only from highest part of the island

☐ Visible from both the area of higher site density and the highest part

4   2   0   4 Kilometers

Fig. 13.7 A comparison of accumulated viewshed from the highest point and the area most densely settled on San Clemente Island

Island suggests that all three forces were in play there: The area on the plateau where sites are found provided a view of those portions of the marine environment that were most accessible and productive. This view was also of many locations on the island itself and of a densely occupied area on Santa Catalina Island, Two Harbors, and intervisibility offered the means by which to coordinate economic and ritual activities. The use of SAR at San Clemente Island enabled us to both very effectively model site patterning, and to develop a surface model with which to better interpret site patterning.

References

Arms, M. (2010). *True north: Journeys into the great northern ocean*. Hinesburg: Upper Access Publishers.

Bongers, J., Arkush, E., & Harrower, M. (2012). Landscapes of death: GIS-based analyses of chalupas in the western Lake Titicaca basin. *Journal of Archaeological Science, 39*, 1687–1693.

Boscana, Gerónomo (1978), Chinigchinich: a revised and annotated version of Alfred Robinson's translation of Father Gerónimo Boscana's historical account of the belief, usages, customs, and extravagancies of the Indians of this mission of San Juan Capistrano called the Acagchemem Tribe. annotations by John P. Harrington; reprinted with a new pref. by William Bright. [Boscana, Gerónimo, 1776–1831]. Banning, Calif.: Malki Museum Press, Morongo Indian Reservation.

Comer, D. C. (2008). Wide-area, planning level archaeological surveys using SAR and multispectral images. In *Proceedings of the geoscience and remote sensing symposium, 2008*. IGARSS 2008, (pp. 45–47). New York: IEEE International.

Comer, D. C., & Blom, R. G. (2007a). Detection and identification of archaeological sites and features using synthetic aperture radar (SAR) data collected from airborne platforms (with Ronald G. Blom). In J. R. Wiseman & F. El-Baz (Eds.), *Remote sensing in archaeology* (Interdisciplinary contributions to archaeology, pp. 103–136). New York: Springer.

Comer, D. C., & Blom, R. G. (2007b). Wide area inventory of archaeological sites using aerial and satellite data sets: Prologue to resource monitoring and preservation (with Ronald G. Blom). In *Proceedings of the 32nd international symposium on remote sensing of environment* (pp. 452–456). Ann Arbor: Environmental Research Institute of Michigan.

Comer, D. C. & Blom, R. G. (2007c). Remote sensing and archaeology: Tracking the course of human history from space (with Ronald G. Blom). *Earth Imaging Journal*, Vol. 2, No. 4 10–13.

Ebert, J. (2000). The state of the art in "inductive" predictive modeling: Seven big mistakes (and lots of smaller ones). In K. L. Wescott & R. J. Brandon (Eds.), *Practical applications of GIS for archaeologists: A predictive modeling toolkit* (pp. 129–134). Taylor and Francis: London.

Fisher, P., & Farrelly, C. (1997). Spatial analysis of visible areas from the bronze age cairns of mull. *Journal of Archaeological Science, 24*, 581–592.

Fogelin, L. (2007). The archaeology of religious ritual. *Annual Review of Anthropology, 36*, 55–71.

Goldberg, C. F. (1993). The application of stable carbon and nitrogen isotope analysis to human dietary reconstruction in prehistoric Southern California. Ph. D. dissertation, University of California, Los Angeles.

Goldberg, C. F., Titus, M., Salls, R., & Berger, R. (2000). Site chronology on San Clemente Island. *Pacific Coast Archaeological Society Quarterly, 36*(1), 31–40.

Kvamme, K. (1988). Development and testing of quantitative models. In W. James Judge & L. Sebastian (Eds.), *Quantifying the present and predicting the past: Theory, method, and application of archaeological predictive modeling* (pp. 325–428). Denver: U.S. Department of the Interior, Bureau of Land Management.

Masters, P. M., Marino, B., & Kennedy, G. (n.d.). Marine paleodiet as indicated by stable nitrogen and carbon iosotopes in burials from the early millingstone horizon. Manuscript in possession of Carol Goldberg.

Meighan, C. W. (2000). Overview of the archaeology of San Clemente Island, California. *Pacific Coast Archaeological Society Quarterly, 36*(1), 1–17.

Porcasi, J., Jones, T. L., & Raab, L. M. (2000). Trans-holocene marine mammal exploitation on San Clemente Island, California: A tragedy of the commons revisited. *Journal of Anthropological Archaeology, 19*(2), 200–220.

Part IV
LiDAR (Light Detection and Ranging)

Chapter 14
LIDAR, Point Clouds, and Their Archaeological Applications

Devin Alan White

Abstract It is common in contemporary archaeological literature, in papers at archaeological conferences, and in grant proposals to see heritage professionals use the term LIDAR to refer to high spatial resolution digital elevation models and the technology used to produce them. The goal of this chapter is to break that association and introduce archaeologists to the world of point clouds, in which LIDAR is only one member of a larger family of techniques to obtain, visualize, and analyze three-dimensional measurements of archaeological features. After describing how point clouds are constructed, there is a brief discussion on the currently available software and analytical techniques designed to make sense of them.

Keywords LIDAR • SAR • Structured light • Point cloud • Photogrammetry • DEM

14.1 Introduction

It is common in contemporary archaeological literature, in papers at archaeological conferences, and in grant proposals to see heritage professionals use the term "LIDAR" (or LiDAR)[1] to refer to high spatial resolution digital elevation models (DEMs) and the technology used to produce them. The goal of this chapter is to break that association and introduce archaeologists to the world of point clouds, in which LIDAR is only one member of a larger family of techniques to obtain, visualize, and analyze three-dimensional measurements of archaeological features. After describing how point clouds are constructed, there is a brief discussion on the currently available software and analytical techniques designed to make sense of them.

---

[1] There is currently no professional consensus with respect to the capitalization of the "I" in the LIDAR acronym, which stands for LIght Detection And Ranging. In this chapter, the established capitalization convention for RADAR (RAdio Detection And Ranging), another active remote sensing technology, is used. The American Society for Photogrammetry and Remote Sensing also follows this convention.

D.C. Comer and M.J. Harrower, *Mapping Archaeological Landscapes from Space*, SpringerBriefs in Archaeology, DOI 10.1007/978-1-4614-6074-9_14, © Springer Science+Business Media New York 2013

14.2   Definition, Construction, and Storage of a Point Cloud

14.2.1   What Is a Point Cloud?

A point cloud is a collection of discrete three-dimensional locations (points) that can have additional metadata associated with each record. By comparison, a DEM is considered to be 2.5D, not 3D, since only one elevation measurement is available for any given cell on a rasterized version of the scanned surface. The most common types of metadata relate to visible color (RGB), intensity, and/or time of collection, all of which can be leveraged to significantly enhance the usability of the cloud.

Point clouds have several unique properties worth mentioning. First, they "look real" to even the most casual observer because of their three-dimensional nature (Fig. 14.1). Second, recorded points can vary greatly with respect to spacing and density. Unlike a DEM, the spatial pattern of points in a cloud may change significantly from one portion of the scanned area to another, resulting in the over-measurement of features in some areas and a complete lack of measurement in others. Third, they allow the researcher to capture complex three-dimensional geometric relationships. Fourth, point clouds can be created in the complete absence of ambient light, which allows for the documentation of archaeological features at night when tourists and/or research personnel are not present.

14.2.2   How Point Clouds Are Constructed

Point clouds can be constructed using a wide variety of technologies and approaches. The following is by no means an exhaustive list, but it is designed to give the reader an overview of the main techniques being used in the field today and a few that are just becoming accessible. They generally fall into one of two categories: (1) active, where the sensor emits energy and uses its interaction with surfaces to construct the cloud and (2) passive, where the sensor collects energy reflected off of surfaces, observed from many different locations, and techniques from the discipline of photogrammetry are used to construct the cloud.

14.2.2.1   Active Scanning Systems

Apart from traditional surveying technologies, active scanning systems are by far the most prevalent ones available to archaeologists and, unfortunately, also tend to be the most expensive because of their reliability, precision, and accuracy. Because they employ technologies that generate their own scanning energy, they can be operated in complete darkness.

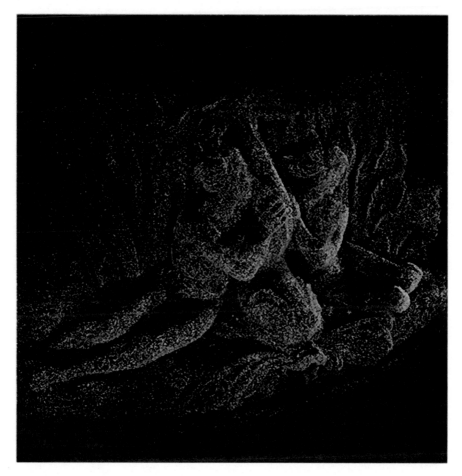

Fig. 14.1 A colorized point cloud representation of a fountain at the Santa Barbara County Courthouse, created by Photosynth™ website user EdLee

LIDAR is the active scanning technique best known to archaeologists, mainly because of its widespread adoption in the surveying community and several recent publications showcasing how it can be used to record, and even discover, archaeological features at both site and landscape scales (Chase this volume, Fisher this volume, Chase et al. 2011; Fisher et al. 2011; Chase et al. 2012). At the most basic level, LIDAR systems send out discrete pulses of light and record both how long it takes those pulses to return and how much of the original energy comes back. That information, when combined with data about where the sensor is positioned and how it is oriented with respect to the real world, is used to construct the point cloud. Each point in the cloud represents a location where the light pulse reflected off of a surface. There are currently four different LIDAR collection modes: linear, flash, full waveform, and Geiger. Each will be explained briefly below.

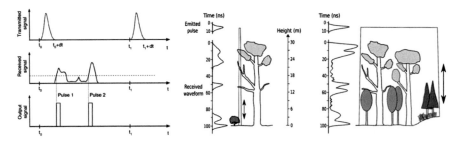

Fig. 14.2 Examples of waveforms. More coarse waveforms (*left*) are converted into discrete returns in traditional linear mode systems. Detailed versions can be recorded using full-waveform systems. Narrow beams (*center*) record more fine-grained details than wide beams (*right*) (Source: Mallet and Bretar 2009, Figs. 2 and 3)

Linear is the most established of the four collection modes. It is so named because a single beam of light is repeatedly projected toward the surface of interest and rapidly moved across it to build up a coherent three-dimensional scan (see Vosselman and Maas 2010 for a more detailed introduction). The closer the sensor is to the features of interest, the more detailed the scan becomes. Once the beam is projected, it may interact with one or more surfaces, resulting in the sensor recording multiple "returns." The first return is usually the surface closest to the sensor, while the last return is usually the one that is farthest from the sensor. It should be noted that the number of returns for any given pulse is determined by analyzing what is known as the waveform, which can be thought of as a continuous line, plotted as a function of time, with one or more peaks that stand out from an otherwise flat background (no surface interactions). The peaks are automatically detected by software connected to the sensor, converted to one or more discrete returns, and linked to specific locations in three-dimensional space with an associated timestamp and an "intensity" value that indicates how reflective the surface is (Fig. 14.2).

Linear mode systems come in two flavors: terrestrial and airborne. Terrestrial systems, usually mounted on tripods, are widely used to support precision surveying activities. Terrestrial systems can and have been used to intensively document individual features and entire sites. Airborne systems are mounted to the bottom of aircraft that fly carefully planned routes over a region of interest to build a complete picture of the landscape. Whereas terrestrial scanners operate from a fixed position, airborne scanners are constantly moving. The position and orientation of the aircraft and the sensor have to be recorded accurately for the LIDAR data to be useful. This requires not only a stable aircraft, but a high fidelity inertial navigation system (INS) and associated software (Fig. 14.3). These requirements have limited the types of aircraft to which LIDAR sensors can be attached (manned fixed wing), but recent industrial advances suggest that it may be possible to mount them on ultralights and unmanned aerial systems.

Flash, the second LIDAR collection mode, can be most easily thought of as an extension of the linear mode. Instead of a single beam of light being projected from the sensor and moved to build up a scan, multiple beams, whose fixed emitters are

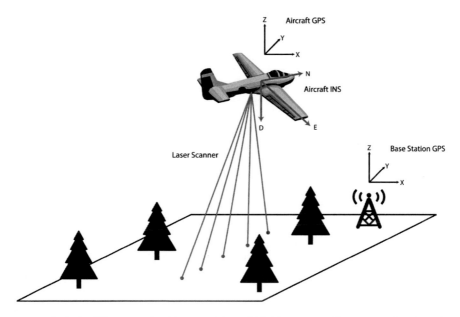

Fig. 14.3 A simplified example of how an airborne LIDAR scanner collects data and registers it to the real world

usually arranged in a two-dimensional array, are projected simultaneously to acquire a scan in one shot that looks like a digital image (c.f. Gelbart et al. 2003; Hanna et al. 2005).

Full waveform, the third collection mode, is similar to the first two in that a beam of light is projected and returns are recorded. What sets it apart, though, is that the beam width is variable and it is designed to collect and store the entire waveform (mentioned above) at a high level of fidelity (Mallet and Bretar 2009). The end result is a waveform that, when oriented vertically with the first surface interaction at the top and the last at the bottom, closely resembles a detailed vertical profile of everything the beam ran into on its way down and back up (Fig. 14.2). Historically, full waveform systems have operated almost exclusively in the airborne domain and have been used for various coarse resolution environmental mapping missions (e.g., NASA's LVIS system), but there is a growing list of examples of its use in archaeology (c.f. Doneus et al. 2008; Lasaponara et al. 2011). One of the main advantages of full waveform LIDAR is that you have a continuous vertical signal that, if necessary, can be broken up into discrete returns to make it compatible with the software and analytical methods discussed below, but can also be used in its native form to conduct more nuanced feature analysis. This emerging technology will have a significant impact on archaeological mapping activities as more systems come online and become accessible to researchers.

The last collection mode is Geiger, which is still highly experimental. This type of sensor, currently limited to airborne applications, uses a very different method of sending and receiving light pulses, relying on avalanche photodiode technology to detect surface returns (Itzler et al. 2011). The photodiodes, generally arranged in a two-dimensional array similar to flash LIDAR, are so sensitive that they can detect a single photon, which means that less energy is needed to form a three-dimensional snapshot. With lower energy requirements, Geiger mode sensors can be flown at higher altitudes than linear mode sensors, resulting in the coverage of larger areas. The downside is that the point clouds produced by Geiger mode systems tend to be very noisy as a result of their increased sensitivity, so producing usable data generally involves a significant amount of post-collection processing using very specialized software.

Synthetic Aperture Radar (SAR) is another active scanning technology that can be used to produce a point cloud. Multiple observations of the same landscape by a spaceborne, airborne, or terrestrial SAR sensor can be fed into an analytical process that results in a very detailed and accurate DEM (Chapman and Blom this volume). What is often not highlighted is that producing the DEM involves converting an irregularly gridded set of derived elevation measurements, with varying spatial density, to a regularly gridded set of measurements so that they can be stored in raster format. In other words, building an elevation model from SAR data begins with building a point cloud.

In addition to higher-profile technologies like LIDAR and SAR, there are a host of desktop and handheld systems currently available or in the process of coming to market. Apart from desktop laser scanners that operate in a similar fashion to terrestrial LIDAR scanners, there are several systems that operate on the principle of structured light. In short, a light pattern is projected onto a surface by a sensor that is either fixed, with the subject of interest rotating in front of it, or handheld, where the user essentially scans a feature by hand. The way that the pattern deforms when it hits a surface allows the associated software to calculate depth and construct a three-dimensional object or scene, either as a solid surface or a point cloud. An interesting recent development has been the repurposing of Microsoft's Xbox 360 Kinect™ sensor (and similar systems like the Leap Motion) into an inexpensive handheld three-dimensional scanning system, which could allow archaeologists to scan artifacts, features, and sites quickly and easily. However, it should be noted that the Kinect is not designed to create archival-quality scans.

14.2.2.2 Passive Scanning Systems

Passive scanning systems can be most easily thought of as standard digital cameras because, in fact, that is exactly what they are. When multiple images of the same scene are captured from different perspectives, the overlapping portions can be used to construct a three-dimensional representation of that scene. This relatively inexpensive technical approach, which is based on well-established photogrammetric principles, is the same as what is used to extract a DEM from overlapping satellite images. First, automated algorithms scan all supplied images in an attempt to find the same distinctive features in as many images as possible. These locations, when combined with the technical specifications of the digital camera(s) used to capture the images, are then fed into another set of algorithms that figure out how to reconcile the three-dimensional locations of these features across all identified images. Once the reconciliation has taken place (known as block adjustment, c.f. Mikhail et al. 2001), additional information correlation algorithms are run to build up a coherent three-dimensional scene.

As with SAR, the precursor to building a raster elevation model from the imagery is an irregularly spaced point cloud that can be treated like any other. More importantly, this feature identification technique can be used with a large number of images to produce an immersive point cloud. Instead of finding just enough interesting features to tie all of the images together, a massive number of them can be located instead. After block adjustment, the three-dimensional locations of these features are known in an arbitrary XYZ space (unless ground control information is provided) and represent a useful point cloud in and of themselves, a concept referred to as Structure from Motion (Hartley and Zisserman 2003).

Microsoft's online Photosynth™ application is a good example of how this approach works. There are imagery-derived point clouds on the Photosynth™ web portal for a large number of archaeological sites (Fig. 14.4), including Stonehenge, but it works for smaller objects, too (Fig. 14.1). While large geospatial software providers like ERDAS and BAE have applications that allow you to extract point clouds from imagery, there are affordable desktop software options as well, including Photomodeler Scanner, PhotoScan, and Bundler. Much like Geiger-mode LIDAR point clouds, though, they can be extremely noisy due to false positive feature matching and transient scene content (e.g., people).

14.3   Exploring and Analyzing Point Clouds

14.3.1   Available Software

Once you have obtained a point cloud, extracting useful information from it can be challenging. Point clouds tend to be massive in size and require specialized software and knowledge to manipulate. Thankfully, the commercial and open source geospatial software communities are actively developing solutions that streamline a wide variety of visualization and analytical tasks. These range from fully-formed desktop applications to lower-level building blocks. On the desktop side (Table 14.1), application functionality, cost, and ease of use vary widely. Many of the applications can visualize a point cloud so that you can interactively explore it; a smaller number have the ability to analyze the cloud directly to extract useful information from it. Most attempt to get around this limitation by first rasterizing the cloud into

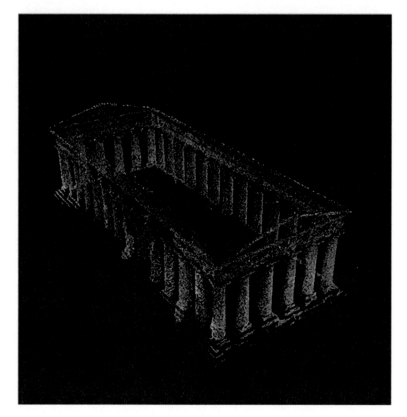

Fig. 14.4 A point cloud of the Doric temple in Segesta, Sicily created by Photosynth™ website user Fewf

a DEM (resulting in loss of data) so that it is compatible with more mainstream 2.5D gridded surface visualization and analysis tools. This is especially true of commercial and open source GIS applications. Outside of the geospatial domain, there are commercial and open source CAD software packages that can handle point cloud data. Cyclone, PolyWorks, RapidForm, GeoMagic, MeshLab, and pointools are good examples.

The available lower-level building blocks, all of which are open source, fall into two categories: data access and data analysis. On the data access side, LibLAS and the Point Data Abstraction Library (PDAL) are extremely useful software libraries that can be integrated with other applications to enable support for point clouds. Paired with this are data analysis libraries like LAStools and Point Cloud Library (PCL).

Table 14.1 Available desktop geospatial software for working with point cloud data

Software	Visualize	Analyze	Rasterize	Free	Easy to use	Notes
Quick TERRAIN MODELER	Yes	Yes	Yes	No	Yes	
Quick TERRAIN VIEWER	Yes	No	No	Yes	Yes	
MARS EXPLORER	Yes	Yes	Yes	No	Yes	
MARS VIEWER	Yes	No	No	Yes	Yes	
QCoherent	Yes	Yes	Yes	No	Yes	Requires ArcGIS
LIDAR ANALYST	Yes	Yes	Yes	No	Yes	Requires ArcGIS
ENVI	Yes	No	Yes	No	Yes	
ArcGIS	No	No	Yes	No	Yes	
GRASS	No	No	Yes	Yes	No	Open source
QGIS	No	No	Yes	Yes	No	Open source
Fugro VIEWER	Yes	No	No	Yes	Yes	
FUSION	Yes	Yes	Yes	Yes	No	

14.3.2 Common Derived Products

While visualizing and interacting with a point cloud can be a useful activity in and of itself, the end goal is usually to extract information that can aid in research and/ or cultural heritage management activities. Analyzing the point cloud directly is hard to do, but significant strides have been made in the domains of surface reconstruction, automatic feature detection and extraction, immersive augmented reality, and multimodal data fusion. Remondino (2011) provides a good overview – albeit mostly at the site level – of what is currently possible, a list of available technologies, and associated challenges.

The more immediately relevant uses of point clouds revolve around established techniques for analyzing raster DEMs. These include derived products such as contours, slope, aspect, hydrology, viewsheds, and shaded relief. For a more complete discussion on raster terrain analysis, the reader is referred to an excellent edited volume by Wilson and Gallant (2000). Since point clouds are almost always collected at a spatial resolution (post spacing) that is significantly higher than the DEMs traditionally available to archaeologists, they open up a new analytical world that can be explored at the human scale.

By far the most popular derived product among archaeologists is the bare earth DEM, also referred to as a Digital Terrain Model (DTM), where everything except the physical surface of the earth has been removed prior to the creation of the raster (Fig. 14.5). This surface is then subjected to a shaded relief analysis in an effort to highlight subtle ground features of archaeological interest (Fig. 14.6, Chase this volume, Fisher this volume, Chase et al. 2011; Fisher et al. 2011; Chase et al. 2012). For many archaeologists, DTM shaded relief output is the only aspect of point clouds they ever see, not knowing that there is a rich world of three-dimensional data sitting behind it. It should be noted that there is no

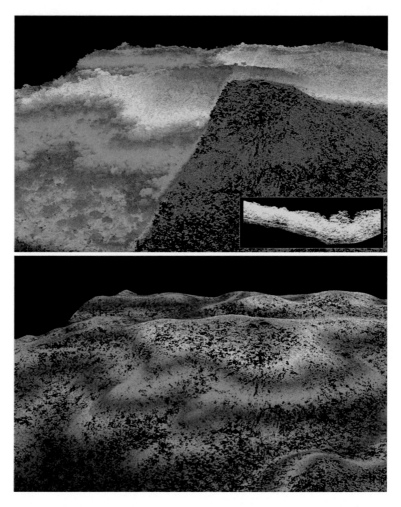

Fig. 14.5 A full point cloud (*top*) for a heavily forested area, depicted using a *blue-green-red* color ramp, and a bare earth point cloud derived from it, depicted in *brown*. The ground is depicted in *pink* in the inset cross-section. The *bottom image* is the bare earth cloud by itself, depicted using the same color ramp (Data courtesy of Arlen Chase, University of Central Florida)

single bare earth generation process and each one has pros and cons (see Vosselman and Maas 2010 for an overview). The best ones might be beyond the financial capabilities of most archaeological projects, but they all work off of the same basic premise: Given enough observations of a densely-covered landscape by an active scanning system, some inevitably come from the ground beneath or next to the cover and can be used in conjunction with an extrapolation process to reconstruct the ground surface. The more ground observations you have, the better the surface reconstruction.

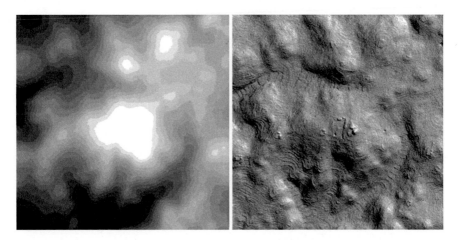

Fig. 14.6 A bare earth DEM (*left*) and its hill-shaded equivalent (*right*). Note the increased visibility of subtle landscape features (Data courtesy of Arlen Chase, University of Central Florida)

14.4 Conclusion

Point clouds, regardless of how they are created, hold a great deal of promise for archaeological research and cultural heritage management – especially as the price of acquisition and processing goes down. At present, they are most widely used in the generation of high spatial resolution DTMs. While the need for 2.5D DTMs will not go away, there is a rich three-dimensional world that is begging to be explored. Software that enables this exploration is still evolving, but there are strong open source and commercial commitments to creating point cloud visualization and analysis tools (and the pace of development is quickening). As stated by Fisher (this volume) and Chase et al. (2012), LIDAR and related technologies are likely to become as integral to the field of archaeology as ^{14}C dating. The sooner this community can embrace the full potential of these technologies, the more it stands to gain.

References

Chase, A. F., Chase, D. Z., Fisher, C. T., Leisz, S. J., & Weishampel, J. F. (2012). Geospatial revolution and remote sensing LiDAR in Mesoamerican archaeology. *PNAS, 109*(32), 12916–12921.

Chase, A. F., Chase, D. Z., Weishampel, J. F., Drake, J. B., Shrestha, R. L., Slatton, K. C., Awe, J. J., & Carter, W. E. (2011). Airborne LiDAR, archaeology, and the ancient Maya landscape at Caracol, Belize. *Journal of Archaeological Science, 38*, 387–398.

Doneus, M., Briese, C., Fera, M., & Janner, M. (2008). Archaeological prospection of forested areas using full-waveform airborne laser scanning. *Journal of Archaeological Science, 35*(4), 882–893.

Fisher, C., Leisz, S., & Outlaw, G. (2011). Lidar – A valuable tool uncovers an ancient city in Mexico. *Photogrammetric Engineering & Remote Sensing, 77*(10), 962–967.

Gelbart, A., Weber, C., Bybee-Driscoll, S., Freeman, J., Fetzer, G. J., Seales, T., McCarley, K. A., & Wright, J. (2003). FLASH lidar data collections in terrestrial and ocean environments. *Proceedings of SPIE, 5086*, 27.

Hanna, B., Chai, B., & Hsu, S. (2005). Wide-area terrain mapping by registration of flash LIDAR imagery. *Proceedings of SPIE, 5791*, 193.

Hartley, R., & Zisserman, A. (2003). *Multiple view geometry in computer vision*. Cambridge: Cambridge University Press.

Itzler, M. A., Owens, E. M., Patel, K., Jiang, X., Slomkowski, K., Slomkowski, K., & Rangwala, S. (2011). *Geiger-mode avalanche photodiode focal plane arrays for 3D LIDAR imaging*. Paper presented at applications of lasers for sensing and free space communications (LSC), Toronto, 10 July.

Lasaponara, R., Coluzzi, R., & Masini, N. (2011). Flights into the past: Full-waveform airborne laser scanning data for archaeological investigation. *Journal of Archaeological Science, 38*, 2061–2070.

Mallet, C., & Bretar, F. (2009). Full-waveform topographic lidar: State-of-the-art. *ISPRS Journal of Photogrammetry and Remote Sensing, 64*(1), 1–16.

Mikhail, E. M., Bethel, J. S., & McGlone, J. C. (2001). *Introduction to modern photogrammetry*. New York: Wiley.

Remondino, F. (2011). Heritage recording and 3D modeling with photogrammetry and 3D scanning. *Remote Sensing, 2011*(3), 1104–1138.

Vosselman, G., & Maas, H. (Eds.). (2010). *Airborne and terrestrial laser scanning*. Boca Raton: CRC Press.

Wilson, J. P., & Gallant, J. C. (Eds.). (2000). *Terrain analysis: Principles and applications*. New York: Wiley.

Chapter 15
The Use of LiDAR at the Maya Site of Caracol, Belize

Arlen F. Chase, Diane Z. Chase, and John F. Weishampel

Abstract With its ability to penetrate dense tropical canopies, LiDAR is revolutionizing how ancient Mesoamerican landscapes are recorded. Locating ancient sites in the Maya area of Central America traditionally employed a variety of techniques, ranging from on-the-ground survey to aerial and satellite imagery. Because of dense vegetation covering most ancient remains, archaeological documentation of the extent of archaeological sites using traditional means was both difficult and usually incomplete. LiDAR was initially applied to the site of Caracol, Belize in April 2009 and yielded a 200 sq km Digital Elevation Model that, for the first time, provided a complete view of how the archaeological remains from a single Maya site – its monumental architecture, roads, residential settlement, and agricultural terraces – were distributed over the landscape. With the detailed information that can be extracted from this technology, LiDAR is significantly changing our perceptions of ancient Maya civilization by demonstrating both its pervasive anthropogenic landscapes and the scale of its urban settlements.

Keywords Geospatial revolution • Landscape archaeology • Mesoamerica • Remote sensing • Urbanism

15.1 Introduction

Tropical and sub-tropical environments provide some of the most challenging conditions for carrying out archaeological research. For the vegetation-enshrouded Maya area of Central America, the task of documenting the anthropogenic landscapes of ancient sites is particularly problematic. Alternating wet and dry seasons found in these areas hasten the decomposition of archaeological remains and encourage the rapid growth of plants and trees over ruins, making them hard to identify and record. Thus, it can be difficult to analyze ancient occupation areas. While several long-term archaeological projects have attempted to extensively map Maya site

D.C. Comer and M.J. Harrower, *Mapping Archaeological Landscapes from Space*, SpringerBriefs in Archaeology, DOI 10.1007/978-1-4614-6074-9_15,
© Springer Science+Business Media New York 2013

cores and surrounding areas, generally only partial samples of the remains from any single site can actually be recorded. Thus, exactly how Maya site cores articulate with surrounding landscapes has been open to interpretation – and the scale of Maya settlements is still a topic of debate. Having worked and mapped in the Maya area for some 40 years, it is clear to us that new technologies can resolve some of the issues facing settlement and landscape archaeology in the Maya region.

Surveying in the Maya area has traditionally proceeded by laborious means, requiring lines-of-sight cut by hand with machetes through the jungle undergrowth followed by the lugging of transits and, more recently, electronic distance meters through the karst topography (Chase 1988). Although less effective underneath the forest canopy, today GPS units make this process slightly easier. The carefully measured points taken with all these instruments are limited both in number and extent and rarely capture all of the smaller features on the landscape, largely because their discovery still requires cutting overgrowth, walking transects, and making on-the-ground measurements. Because heavy vegetation generally leads to sampling rather than to the full documentation of sites through survey, the extent and scale of Maya urban settlements and landscape manipulation has been difficult both to establish and to visualize.

Until recently, mapping at Caracol, Belize followed methods that had been utilized by Mayanists for more than a century, resulting in a traditional, 23 sq km rectified plan. Maya ruins are characterized by stone and earth construction of ancient buildings, platforms, reservoirs, and roadways. In most cases, these ancient architectural remains are raised but have lost their form overtime through the effects of vegetation and natural weathering. Both well-constructed stone buildings and the perishable buildings that once formed ancient residential groups appear today as mounds of variable height; roadways are similarly raised above the landscape. In contrast, constructed reservoirs appear as roughly rectangular-shaped depressions into the surface of the land. All of these earth-covered remains are relatively easily recognized and mapped (e.g., Fig. 15.1), if one has the requisite time.

At Caracol, however, the landscape was also covered with lines of stone, recognized as the remains of ancient agricultural terraces, that ranged up to 3 m in height and that appeared in regularized patterns over the site's valleys and hills. Had the site been small, the 23 sq km of area surveyed between 1983 and 2003 might have sufficed to define the city. However, the sampling transects that were cut into the surrounding countryside uncovered Maya settlement well outside of the site epicenter with no clear areas of settlement drop-off. Road systems leading out from the Caracol epicenter were linked to other monumental architecture and plazas some 3–8 km distant; all indications were that residential settlement and agricultural terraces were continuous between these nodes (Chase and Chase 2001). What emerged from conventional mapping at Caracol was a huge urban settlement that did not align well with preconceptions of Maya society based on other data classes. Maya hieroglyphs had been used to reconstruct a landscape populated with numerous small competing polities, each with a royal court (see Chase and Chase 2008) and ancient farmers were still viewed as largely practicing extensive swidden agriculture rather than focusing on more intensive systems.

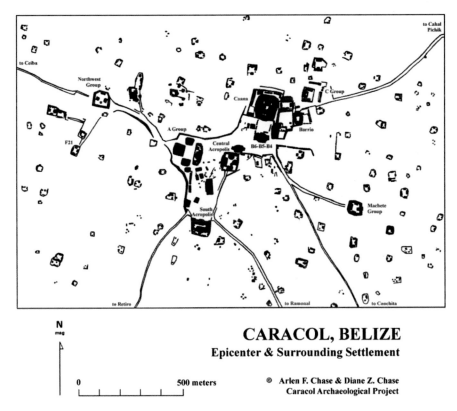

Fig. 15.1 Traditional rectified archaeological map of structural remains in central Caracol created through on-the-ground survey

However, labor-intensive block mapping could not record all of the agricultural terraces that were present at Caracol – simply because there were so many of them and the vegetated understory was too dense to easily accomplish this goal (Chase and Chase 1998). And, neither the full extent of the settlement nor the point of settlement drop-off could be discerned. To remedy this situation and to better understand the totality of the site's settlement, we turned to technology in the form of airborne LiDAR. This geo-spatial technology had not been previously or effectively used at this scale or point density anywhere in Central or South America before this application. The use of LiDAR at Caracol – and subsequently elsewhere in tropical environments – is revolutionizing landscape archaeology and our spatial understanding of past societies (Chase et al. 2012).

15.2   Light Detection and Ranging (LiDAR)

In the early part of the twenty-first century, innovative site documentation efforts were focused on satellite imagery and airborne or satellite-borne radar (Wiseman and El-Baz 2007); however, none of these technologies have the resolution that is necessary to fully identify and document the extent of ancient landscape modifications, especially beneath tree cover. Vegetation covering the ancient remains often covers all but monumental architecture. Thus, neither the identification of the entirety of ancient occupation nor the determination and mapping of architectural forms is possible using these technologies. One of our interdisciplinary research group used LiDAR to image the forest canopies of Costa Rica (Weishampel et al. 2000) and was familiar with the technology and its possibilities. LiDAR had not been previously pursued in the Maya area because earlier archaeological tests of this technology in Costa Rica were not successful (Sheets and Sever 1988).

However, after Hurricane Mitch in Honduras in 1998, LiDAR was flown over coastal areas to assess the damage and a LiDAR image of Copan was published (Gutierrez et al. 2001); although not perfect and aided by undergrowth that had been artificially thinned in the Copan Park, it appeared to demonstrate the potential of this technology to see beneath a jungle canopy. Initial publications also indicated LiDAR's ability to penetrate European forests (Devereux et al. 2005). While less dense than the Maya sub-tropical forest cover, the lasers employed in the European aerial application of LiDAR passed through canopies, returning detailed bare earth information. Thus, given advances in technology, LiDAR presented a potentially viable solution to the issues of scale and visibility that were confronting Maya archaeology.

15.3   The Caracol LiDAR Application

The first large-scale application of airborne LiDAR in Mesoamerica was undertaken at Caracol, Belize in April of 2009 (Chase et al. 2010, 2011; Weishampel et al. 2010). With funding from NASA [NNX08AM11G] and the UCF-UF Space Research Initiative, a 200 sq km area of west-central Belize was overflown by the National Center for Airborne Laser Mapping (NCALM) in a gridded pattern at an elevation of 800 m over the course of 5 days. The sensor was an Optech GEMINI Airborne Laser Terrain Mapper (ALTM) that was flown aboard a twin-engine Cessna Skymaster. The campaign required 25.4 h of flight time and 9.2 h of laser-on time to survey the region. The discrepancy was due to occasional cloud cover below the airplane, which flew nominally 800 m above ground level.

Two billion, three hundred and eighty million laser pulses were fired, resulting in 4.28 billion measurements that constitute a 3D (x,y,z) "point cloud" with approximately 20 points per square meter. Of these, 1.35 points per sq m on average were classified as ground points, yielding a conservatively estimated vertical resolution of 5–30 cm. The ground point density varied as a function of vegetative cover from a few ground points in a 10 by 10 m area for a dense canopy to more than 1,000 points in a comparable treeless area. The classification of ground points is accomplished through a computer intensive process of iteratively building triangulated surface models based on the lowest returns. It involved removing outliers and correlating the measures to elevations of greater than 1,600 check points (i.e., known ground locations which had been previously measured).

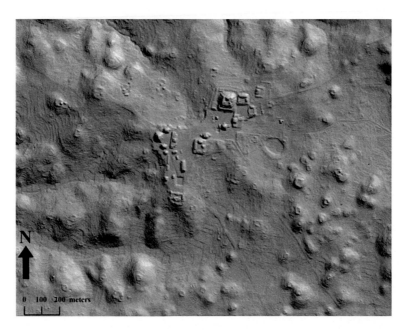

Fig. 15.2 LiDAR bare-earth visualization in 2D of epicentral Caracol. Terraces seen in this figure are not shown in Fig. 15.1 and were not surveyed on the ground due to the difficulty in undertaking traditional mapping in the karst terrain. Approximately 12% of the residential groups that can be identified in this LiDAR image were not recovered through traditional survey methods and some of the mapped groups were not correctly located

The resulting point cloud is the first time that a large number of Maya constructions have been successfully viewed through the encompassing jungle canopy (Fig. 15.2). It is also the first time that the full extent of an ancient Maya city has been visualized in terms of its landscape (Fig. 15.3). For Caracol, the LiDAR analysis conclusively demonstrates that the areal extent of the city was more than 180 sq km. The LiDAR analysis not only recorded thousands of individual residential structures and groups on flat ground, but also recorded almost 5,000 elevated residential groups situated within more than 160 sq km of continuous terracing. Both known and newly discovered causeways that linked outlying monumental plazas to the site epicenter were also easily seen in the Digital Elevation Model (DEM) that was generated from the point cloud data.

Thus, an entire city in the Southern Maya lowlands can be "seen" and its scale appreciated, thus obviating the need for speculation based on a partial sample. Previous investigations serve to "ground-truth" the LiDAR and provide information on time depth and functions for human-made features. Because of 30 years of archaeological research, we know that the settlement visible in the LiDAR-generated DEM dates mostly to the city's Late Classic Period (A.D. 550–900) peak of occupation. At this time, the Caracol settlement was continuous over this landscape and the nodes of larger monumental plazas and architecture (Fig. 15.4) were clearly integrated into a single urban system through the site's causeways (Chase and Chase 2013).

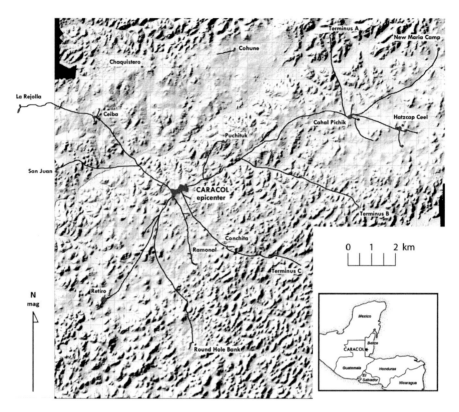

Fig. 15.3 Two hundred sq km hillshaded DEM of Caracol, Belize with the site's road system and termini highlighted. Approximately half the causeways were not recorded through traditional survey. At this reduced scale, no terraces, residential groups, reservoirs, or public architecture are visible; however, most of the landscape is covered with residential groups and agricultural terracing

The LiDAR data from Caracol shed light on a number of issues and data classes related to both ancient and modern agricultural practices. First and foremost, they fully reveal the extent of not only settlement but also agricultural terracing (Fig. 15.5). The LiDAR demonstrates that much of the landscape – some 80% of the 200 sq km DEM – is covered with these terraces, both hillsides and valleys, showing how the Maya of Caracol maintained their large population numbers and size. The vertical control obtainable in the LiDAR data is also sufficiently discriminating to indicate how the terraces changed water flow. Not only do terraces retain water, but also their construction on the landscape created a complex form of water management that largely has precluded erosion of the landscape. Even 1,000 years after being abandoned, the terraces still effectively manage the flow of water on the Caracol landscape. The continued ability of the terraces to retain water has an effect on the height of the canopy; trees on the anthropogenic terraces are approximately 2 m higher than their counterparts that do not occupy these favorable locations (Hightower 2012). The species composition found

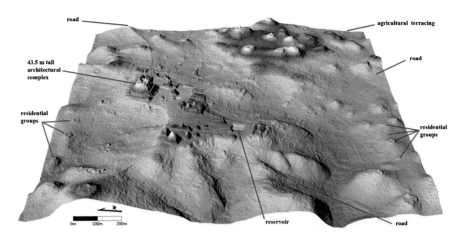

Fig. 15.4 LiDAR bare-earth visualization in 2.5D of central Caracol. Monumental architecture, reservoirs, roads, and residential groups are all visible; some agricultural terraces may be seen in this visualization, but many more are actually present and visible when images are viewed at a larger scale and different hillshade

on the anthropogenic landscape also is significantly different than that found in areas that have not been modified by human agency (Hightower 2012), showing the impact of ancient actions on modern vegetation. LiDAR also provides researchers with a better way to measure the illegal deforestation related to modern agriculture that is currently taking place along Belize's border with Guatemala in the Caracol Archaeological Reserve (Weishampel et al. 2012).

LiDAR elevation data further permit the identification both of deeper sinks/caves beneath the canopy and of shallower depressions that are representative of anciently constructed Maya reservoirs. Some 61 karst depressions or cave openings have been identified within the 200 sq km DEM (Weishampel et al. 2011). The point cloud data provide an uncanny ability to both accurately model the shape of these features and to provide illustrations of the overlying canopy that obscures them. More difficult to identify are the many constructed reservoirs that dot the Caracol landscape. Some ancient reservoirs are quite sizeable (Fig. 15.6), but the vast majority are quite small. However, even these are visible in the point cloud data (as are 1 m wide chultun entrances and looted tombs). Visual inspection of the hill-shaded bare-earth DEMs resulted in the identification of some 271 reservoirs; edge-detection methods, which identify linear features on the landscape, have been applied to portions of the DEM. The results suggest that some 1,400 reservoirs exist in the LiDAR data (Chase 2012). These data have implications for the interpretation of water control by the ancient Maya, suggesting that residential groups controlled their own water sources outside the purview of the Maya elite (Chase 2012).

Perhaps the most significant impact of LiDAR on our view of the Maya landscape is that we can no longer conceive of Maya cities as being small individual "dots" on a map. LiDAR conclusively shows that at least some sites occupied large

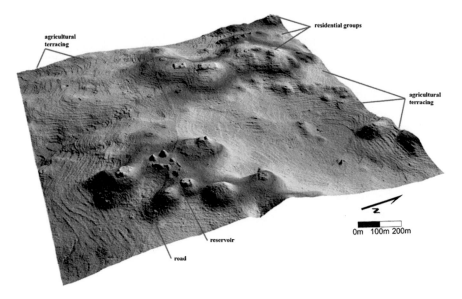

Fig. 15.5 LiDAR bare-earth visualization in 2.5D of Caracol's Ceiba terminus, showing the extent of the terraced landscape. While public architecture, roads, and agricultural terraces are readily visible in this visualization, many of the residential groups cannot be seen due to the reduced scale and hillshade used

areas, in Caracol's case approximately 180 sq km. The size and concentration of this urban settlement is consistent with a form of low-density urbanism found in other tropical parts of the ancient world (Fletcher 2012) and is helping to remove a former theoretical bias among researchers that held that civilizational development was limited within tropical settings. However, it will take more LiDAR covering even larger spatial areas to answer other key questions about ancient Maya polities: How large were they? And, what did borders, boundaries, and frontiers between Maya polities looks like? LiDAR provides a tool that enhances our ability to undertake large-scale spatial research.

15.4   Significance

The successful application of LiDAR at Caracol has had a major impact on Mesoamerican archaeology in that it has helped demonstrate the scale and complexity of ancient Maya city and landscape organization (Chase et al. 2012). For the first time, LiDAR has permitted researchers to view the scale of landscape modification and settlement distribution by the Maya without having to revert to arguments over whether or not the recorded sample is representative. By providing a comprehensive view of the ancient settlement and its distribution over a landscape, LiDAR data are changing forever the ways in which Mesoamerican sites can be mapped and recorded. The full

Fig. 15.6 Caracol's "A Group" reservoir (also visible in Fig. 15.4), showing a vertical cross- section of canopy over the reservoir (*top*), a 2D LiDAR bare-earth visualization showing both the reservoir and the position of the cross-section (*middle*); and a photograph of the reservoir looking south (*bottom*)

landscape coverage provided by LiDAR overshadows the data gained from traditional mapping, leading to fuller and more detailed interpretations. Whereas multiple models could compete when societal interpretations were based on limited archaeological sampling of the landscape (as it was with traditional mapping), the totality of the LiDAR data can also provide boundary conditions for the theoretical perspectives that can be correctly applied to ancient Maya civilization during the Classic Period (Chase et al. 2011). When LiDAR can be conjoined with detailed archaeological data, as it can at Caracol, a richer and more nuanced view of the Maya past is gained.

LiDAR is at the forefront of the geospatial revolution that is sweeping through Mesoamerican archaeology. It is changing the way in which landscapes and environments are perceived and will alter our interpretations concerning past human-nature relationships. Ancient Maya societies were not uniform and, as more LiDAR is obtained, the spatial variability that once existed will become better defined. Significantly, however, LiDAR provides the ability to gain a larger, more representative, and detailed sample of a site or a region, providing solid data sets upon which to base further interpretational refinements concerning the scale and integration of ancient Maya land use and manipulation.

Acknowledgements The LiDAR research undertaken at Caracol was supported by NASA Grant #NNX08AM11G through the Space Archaeology program and a University of Central Florida – University of Florida – Space Research Initiative (UCF-UF-SRI) grant. We appreciate the support of the Belize Institute of Archaeology in carrying out this work and also the ongoing assistance from NCALM scientists Ramesh Shrestha, William Carter, and Michael Sartori. The archaeological research at Caracol has been supported by: the Ahau Foundation; the Alphawood Foundation; the Dart Foundation; the Foundation for the Advancement of Mesoamerican Studies, Inc.; the Geraldine and Emory Ford Foundation; the Government of Belize; the Harry Frank Guggenheim Foundation; the National Science Foundation [grants BNS-8619996, SBR-9311773, and SBR 97–08637]; the Stans Foundation; the United States Agency for International Development; the University of Central Florida; and, private donations. The authors also wish to acknowledge the helpful comments of three reviewers.

References

Chase, A. F. (1988). Jungle surveying: Mapping the archaeological site of Caracol, Belize. *P.O.B. (Point of Beginning), 13*(3), 10–24.

Chase, A. S. Z. (2012). *Beyond elite control: Maya water management at the site of Caracol, Belize.* Senior Thesis, Departments of Archaeology and Computer Science. Cambridge, MA: Harvard University. http://www.caracol.org/include/files/chase/asz12.pdf. Accessed 10 Aug 2012.

Chase, A. F., & Chase, D. Z. (1998). Scale and intensity in classic period Maya agriculture: Terracing and settlement at the 'Garden City' of Caracol, Belize. *Culture and Agriculture, 20*(2), 60–77.

Chase, A. F., & Chase, D. Z. (2001). Ancient Maya causeways and site organization at Caracol, Belize. *Ancient Mesoamerica, 12*(2), 273–281.

Chase, D. Z., & Chase, A. F. (2008). Que no nos Cuentan los Jeroglificos?: Arqueología e Historia en Caracol, Belice. *Mayab, 20*, 93–108.

Chase, D. Z., & Chase, A. F. (2013). Low-density urbanism at Caracol, Belize: Centralization and control in a sustainable landscape. In L. Lucero & R. Fletcher (Eds.), *Low density urbanism in the tropical landscape: Sustainability and transformation.* Tucson: University of Arizona Press (in process).

Chase, D. Z., Chase, A. F., Awe, J. J., Walker, J. H., & Weishampel, J. F. (2011a). Airborne LiDAR at Caracol, Belize and the interpretation of ancient Maya society and landscapes. *Research Reports in Belizean Archaeology, 8*, 61–73.

Chase, A. F., Chase, D. Z., Fisher, C. T., Leisz, S. J., & Weishampel, J. F. (2012). Geospatial revolution and remote sensing LiDAR in Mesoamerican archaeology. *Proceedings of the National Academy of Sciences, 109*(32), 12916–12921. doi:10.1073/pnas.1205198109.

Chase, A. F., Chase, D. Z., & Weishampel, J. F. (2010). Lasers in the jungle: Airborne sensors reveal a vast Maya landscape. *Archaeology, 63*(4), 27–29.

Chase, A. F., Chase, D. Z., Weishampel, J. F., Drake, J. B., Shrestha, R. L., et al. (2011b). Airborne LiDAR, archaeology, and the ancient Maya landscape at Caracol, Belize. *Journal of Archaeological Science, 38*, 387–398. doi:10.1016/j.jas.2010.09018.

Devereux, B. J., Amable, G. S., Crow, P., & Cliff, A. D. (2005). The potential of airborne lidar for detection of archaeological features under woodland canopies. *Antiquity, 79*, 648–660.

Fletcher, R. (2012). Low-density, agrarian-based urbanism: Scale, power, and ecology. In M. Smith (Ed.), *The comparative archaeology of complex societies* (pp. 285–320). New York: Cambridge University Press.

Gutierrez, R., Gibeaut, J. C., Smyth, R. C., Hepner, T. L., & Andrews, J. R. (2001). Precise airborne lidar surveying for coastal research and geohazards applications. *International Archives of Photogrammetry and Remote Sensing, 34*(3), 185–192.

Hightower, J. N. (2012). *Relating ancient Maya land use legacies to the contemporary forest of Caracol, Belize*. M.A. Thesis, Department of Biology. Orlando: University of Central Florida. http://www.caracol.org/include/files/chase/HightowerMA2012.pdf. Accessed 10 Aug 2012.

Sheets, P. D., & Sever, T. (1988). High tech wizardry. *Archaeology, 41*, 28–35.

Weishampel, J. F., Blair, J. B., Knox, R. G., Dubayah, R., & Clark, D. B. (2000). Volumetric lidar return patterns from an old-growth tropical rainforest canopy. *International Journal of Remote Sensing, 21*, 409–415.

Weishampel, J. F., Chase, A. F., Chase, D. Z., Drake, J. B., & Shrestha, R. L., et al. (2010). Remote sensing of ancient Maya land use features at Caracol, Belize related to Tropical Rainforest Structure. In S. Campana, M. Forte, & C. Liuzza (Eds.), *Space, time, place: Third international conference on remote sensing in archaeology* (pp. 45–52). British Archaeological Reports S2118. Oxford: Archaeopress.

Weishampel, J. F., Hightower, J. N., Chase, A. F., & Chase, D. Z. (2012). Use of airborne LiDAR to delineate canopy degradation and encroachment along the Guatemala-Belize border. *Tropical Conservation Science, 5*(1), 12–24.

Weishampel, J. F., Hightower, J. N., Chase, A. F., Chase, D. Z., & Patrick, R. A. (2011). Detection and morphologic analysis of potential below-canopy cave openings in the karst landscape around the Maya polity of Caracol using airborne LiDAR. *Journal of Cave and Karst Studies, 73*, 187–196.

Wiseman, J., & El-Baz, F. (2007). *Remote sensing in archaeology*. New York: Springer.

Chapter 16
New Perspectives on Purépecha Urbanism Through the Use of LiDAR at the Site of Angamuco, Mexico

Christopher T. Fisher and Stephen J. Leisz

Abstract Advances in LiDAR technology promise to change the way that ancient architectural remains are documented, analyzed, and managed at Mesoamerican urban centers. Here we discuss the way that LiDAR has helped document the location, temporal associations, and spatial arrangement of ancient architecture at the Purépecha city of Angamuco, located within the Lake Pátzcuaro Basin, Michoacán, Mexico. Angamuco occupies a rugged topographic feature that has served to preserve ancient architectural features to a degree not typically seen within the region. As a supplement to full-coverage survey we obtained dense LiDAR data for 9 sq km of the settlement that clearly show over 20,000 architectural features from the urban core of the city. Through the use of LiDAR we were able to more quickly and accurately determine the size of the ancient city, better document the type and distribution of ancient features, and significantly change the manner in which we conducted the survey.

Keywords Mesoamerica • Mexico • Ancient urbanism • LiDAR • Archaeology • Angamuco • Ancient architecture

16.1 Introduction

The variation and arrangement of architecture in ancient cities creates distinctive cultural markers, and an understanding of this spatial variation can provide powerful examples for modern social organization (Feinman 1991; Fisher et al. 2009; Sabloff 2008; Sabloff and Ashmore 2001; Smith 2006, 2010b). Archaeological documentation of the urban form, however, typically involves years of painstaking survey, mapping, and geospatial analysis. Aerial-based LiDAR promises to significantly advance archaeological understanding of complex urban forms in forested environments such as those that cover much of Mesoamerica (Chase et al. 2011, 2012, and this volume).

D.C. Comer and M.J. Harrower, *Mapping Archaeological Landscapes from Space*, SpringerBriefs in Archaeology, DOI 10.1007/978-1-4614-6074-9_16, © Springer Science+Business Media New York 2013

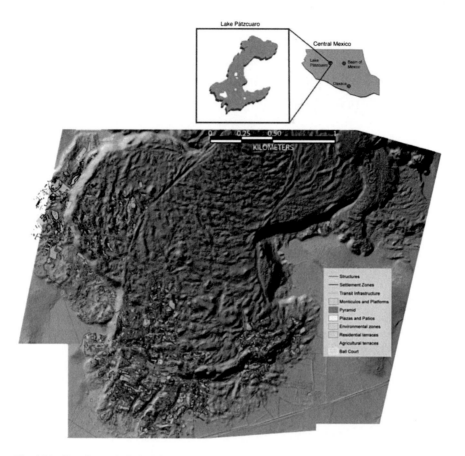

Fig. 16.1 *Top*; shows the Lake Pátzcuaro Basin within Central Mexico. *Bottom*; hillshade based on a 25 cm digital elevation model derived from LiDAR data showing the Angamuco malpaís overlain with the 7,000 digitized architectural features mapped during survey. When enlarged to a 6′ × 6′ map the image shows over 20,000 human generated features that completely cover this landform

Here we discuss the way that LiDAR has helped document the location, temporal associations, and spatial arrangement of ancient architecture at the Purépecha city of Angamuco, located within the Lake Pátzcuaro Basin, Michoacán, Mexico (LPB) (Fig. 16.1, top). At the time of European contact the LPB was the core of the Late Postclassic (LPC) (A.D. 1350–1520) Purépecha Empire. Cities were an important component of Purépecha statecraft but comparatively little is known about their general characteristics, organization, and evolution (Beekman 2010; Michelet 2008; Pollard 2008). For Angamuco, LiDAR analysis both enriched traditional survey results while at the same time allowing for the rapid documentation and classification of buildings outside areas of survey coverage. In addition, LiDAR analysis aided in the identification and configurations of buildings representing spatial divisions within the city. From these data we have been able to create models for the evolution of the city that are not normally possible without years of research.

16.2   Background

We follow the premise that the form and arrangement of ancient architecture held social significance that can yield important insights into ancient socio-political and socio-economic organization (Joyce 2009; Marcus and Sabloff 2008; Smith 2010a: 137; Smith 2003). High-resolution geospatial technologies are transformative in this regard because they can record cultural patrimony in multiple dimensions with great accuracy over wide areas (Chase et al. 2012). These three-dimensional documents can be used to investigate not only archaeological settlements but also their integrated hinterlands at scales ranging from individual buildings, cities, and engineered environments (Fisher et al. 2011). These records have great utility in and around ancient urban environments serving as indelible three-dimensional documents that record everything between the earth's surface and the upper limits of the floral/faunal canopy. As a consequence, aerial-based LiDAR is poised to revolutionize the study of urban settlements in Mesoamerican archaeology (see Chase et al. 2012, this volume, White, this volume).

One such example is the Mesoamerican city of Angamuco, located within the LPB, an area of cultural and historical significance that has long been recognized as a Mesoamerican core area. Over the last 4 years, we have been conducting full-coverage archaeological survey within the region to better understand socio-economic transformations associated with the development of the Empire including changes in settlement systems during the Postclassic (A.D. 1,000–1,520), and the nature and organization of the ancient built environment (Fisher 2009, 2010, 2011). During the course of this research, our team began surveying a settlement that we quickly realized was larger, more spatially complex, and predominately earlier then suggested by current models (e.g. Pollard 2008).

16.3   Angamuco

This urban center, named Angamuco, occupies a mid-Holocene lava flow (malpaís) at the eastern boundary of the LPB (Fig. 16.1, bottom). Rugged topography and closed forest canopy covering the site limit modern land-use. This has preserved Prehispanic architectural foundations and other features that are readily visible on the surface including thousands of building foundations, platforms, pyramids, roads, terraces, granaries, and walls, manufactured from stacked stone with undisturbed floors, middens, and other deposits (Fig. 16.1, bottom). The size, complexity, and spatial organization of Angamuco are beyond what can arguably be characterized as a middle range polity and is therefore best recognized as a city (papers in Marcus and Sabloff 2008).

Over the course of 2 years (2009–2010) we conducted the full-coverage pedestrian survey of just over 2 sq km of the ancient city following traditional methodologies coupled with sub-meter GPS/handheld computers that allowed us to document more than 2,500 architectural and landscape features (Fisher 2011) (Fig. 16.1, bottom). Artifact surface collections provided information on the age of structures and the nature of their occupations (see Fisher 2010, 2011).

The architecture at Angamuco occurs as standardized forms that re-occur over much of the malpaís, allowing us to create a comprehensive 60-point architectural typology detailed in Fisher 2010 (and forthcoming publications). This is similar to the ideal types of the Gulf Lowlands (papers in Stark and Arnold 1997; Stark 1999), patio groups of the Maya lowlands (e.g. Ashmore 1981), and other Mesoamerican sites such as Xochichalco (Hirth 2000) and Teotihuacan (Cowgill 2004). Combinations and forms of the individual elements varies by factors such as public/domestic, elite and commoner, house/platform size, monumentality, access to public and private spaces, proximity to major roadways, and proximity to site entrances.

Reconnaissance conducted in 2010 outside areas of full-coverage survey documented the presence of architectural features that covered at least 12 sq km of the malpaís though the spatial continuity, form, and arrangement of these features was unknown. Given the rugged topography documenting these additional areas through intensive methodologies would have taken what we estimated to be at least a decade of fieldwork and analysis. In an effort to hasten our understanding of Angamuco in 2010 we obtained aerial-based LiDAR for 9 sq km of the Angamuco urban area (Fisher et al. 2011) (Appendix 1). The LiDAR coverage reveals more than 20,000 buildings and other landscape features that completely cover the entire area of coverage (Fig. 16.1, bottom). In terms of accuracy we were able to directly overlay previously surveyed features onto LiDAR derived products in a manner that is identical to that described by Chase et al. in this volume and shown in Fig. 16.2. The LiDAR analysis also demonstrates a repetition of the basic patterns that we documented during full coverage survey including building types, spatial arrangements, associations, and a completely humanized landscape. Figure 16.3 shows two examples of color shaded visualizations of one such densely settled area of Angamuco. In the center of this figure is a large flattened plaza with two central altars, flanked by a pyramid on the left side and an elite complex on the right. These zones are encircled by residential neighborhoods demarcated by walled roads and passages. The three views on the bottom of Fig. 16.3 show LiDAR derived profiles of this same area.

To further field-check the LiDAR results, we conducted a season of full-coverage survey in 2011 using LiDAR derived products such as hillshades and high-resolution digital elevation models as a base layer onto which we digitized ancient buildings in real time. Through this work we were able to confirm the accuracy of the newly acquired LiDAR data while developing new techniques to integrate these products into our survey methodologies. The addition of these 2D + data allowed a doubling in the amount of area covered and the number of architectural features documented from an average of 1 sq km/1,400 buildings to almost 3 sq km/4,000 buildings per season. Importantly, with these new methodologies, we believe that it would be possible to survey much of the site in just over 6 months of intensive fieldwork. This represents an incalculable savings of effort and money that can be redirected toward other avenues of investigation at Angamuco.

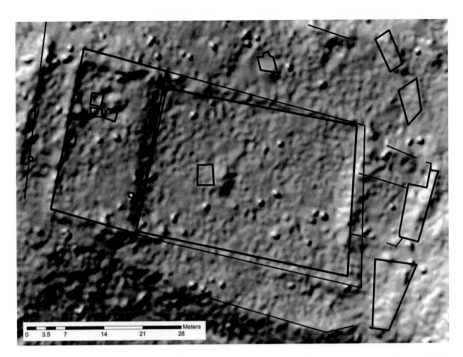

Fig. 16.2 Hillshade shown in grey based on a 25 cm digital elevation model derived from LiDAR data overlain with previously surveyed archaeological features showing the accuracy and detail of the LiDAR analysis

Based on our three seasons of full-coverage survey aided by LiDAR analysis, we have been able to reach several conclusions regarding the nature of the Angamuco occupation that would have normally not been possible without decades of intensive fieldwork. First, we can demonstrate that Angamuco was occupied from at least the Late Classic to Contact periods (A.D. 900–A.D. 1520) in three phases with distinct architectural associations.

The sequence begins in the Epiclassic-Early Postclassic periods (EPC) (A.D. 900–1200) with the founding of the city and a focus on sunken patio complexes that are similar to those from the Bajio region of Mexico (Cárdenas García 1999; Darras and Faugère 2005). This is followed by an episode of major growth and expansion during the Middle Postclassic (MPC) (A.D. 1200–1350) centered around several nodes, each with distinct rectilinear pyramid complexes that are similar to those documented for the adjacent Zacapu Basin (Arnauld and Faugère-Kalfon 1998; Michelet 1996, 2000, 2008).

The final phase sees a contraction of the settlement area during the LPC (A.D. 1350–1520) with a focus around at least two nodes with Purépecha Imperial style architecture composed of circular and square elements (Fig. 16.2), social differentiation, and increased variety in architectural form. This phase is similar to other small LPC centers within the LPB (Fisher 2005, 2007, 2009; Fisher et al. 2003; Pollard 2008).

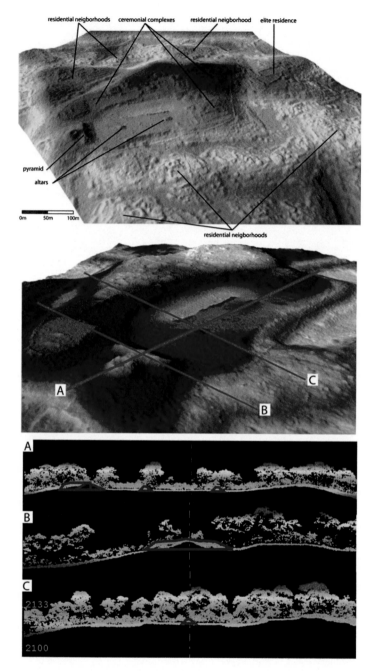

Fig. 16.3 Three dimensional visualizations of a civic ceremonial zone embedded within an elite neighborhood. *Top*, color shaded view based on a 25 cm digital elevation model derived from LiDAR; *Middle*, three dimensional view of the main platform from this zone; *Bottom*, profiles derived from LiDAR data of the main platform, letters in *middle* show the location of transects, shapes in *red* represent the pyramid and altars

16.4 Spatial Complexity at Angamuco

We can also identify spatial divisions within the city that we believe have social significance and that we will begin testing through excavation in 2013. Natural and human-generated divisions form a nested hierarchy of four delimited zones with a clear inside/outside dichotomy that we presume had ethnic, familial, status, or other associations to the ancient inhabitants (Fig. 16.4). The first of these is the functional unit and represents the basic "building block" of Angamuco social space. The functional unit was the locus of daily activities and served as the primary residence or dwelling. Spatially, a functional unit is represented by a single, or a small series, of related buildings, platforms, mounds, or other features that are often marked by a combination of walls, linear mounds, and roads.

Angamuco functional units are arranged into complejos of similar composition that vary significantly in time and space (Fig. 16.4, top). Complejos represent a mid-point between a single functional unit and a neighborhood. Unlike a neighborhood, however, complejos represent only one or two functions. Additionally, not all neighborhoods need be composed of complejos. Our working hypothesis is that complejos first appeared during the EPC as "plaza group" complexes (e.g. Ashmore 1981) composed of people related by kinship, socio-economic status, or occupation. Most commonly they are composed of a series of platforms and terraces arranged around a central, shared plaza. These are similar to those documented for Mesoamerica and that have recently been examined at Angamuco by Bush (see summary and discussion 2011).

Complejos and functional units are arranged into neighborhoods (Fig. 16.4, middle). Here we follow Smith (2010a: 139) who defines a neighborhood as "a residential zone that has considerable face-to-face interaction and is distinctive on the basis of physical and/or social characteristics." Importantly, the spatial groupings that we define as neighborhoods at Angamuco typically include many types of architecture such as civic-ceremonial, storage, and boundary forms. Thus, neighborhoods at Angamuco are not solely residential but instead are multifunctional and designed to serve the needs of *the neighborhood*.

Many Angamuco neighborhoods also contain architecture that we have preliminarily assigned to more than one social class, meaning that elite and commoner forms of architecture are present in a single neighborhood. Deciphering the relationship between these social classes provides a significant focus of future research. It is tempting to propose that neighborhoods represent something akin to *calpulli*-like forms (or the Altepetl) as mentioned in the *Relación de Michoacán* (Pollard 1993), but much more research is needed. This trend seems to intensify in the LPC and possibly signals an increase in social differentiation.

Walls, platforms, terraces, and natural boundaries commonly demarcate neighborhoods and serve to delimit public areas. Some architectural features may have supported large screens or fences as mentioned in the *Relación de Michoacán* (1542). Neighborhoods also tend to have centralized public space operationalized as large plazas with entrances, altars, and other public features.

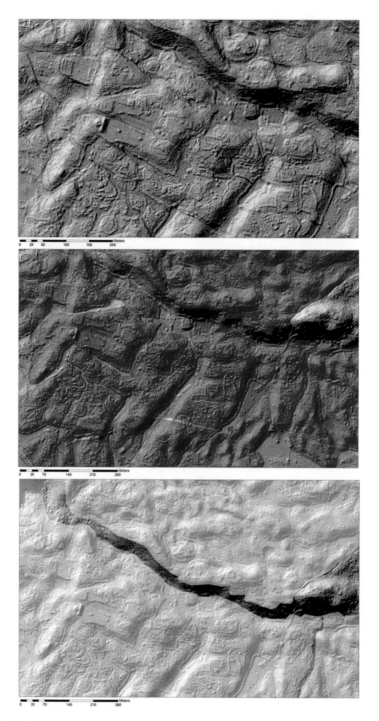

Fig. 16.4 Views of complejos (*top yellow*), neighborhoods (*middle purple*), districts (*bottom white*) for the same zone shown in Fig. 16.3

Neighborhoods tend to have formalized ways of traversing public and private spaces. Roads with entrances marked by mounds and platforms are commonly found at neighborhood boundaries. These same entrances may have also supported gates, again mentioned in the *Relación de Michoacán*. Additionally, networks of passages mark formal ways to move through space that are internally contained within neighborhood boundary features.

The largest spatial zone that we can identify through material traces on the landscape can best be characterized as a ward or district following Smith (2010a) (Fig. 16.4, bottom). Districts at Angamuco are typically multi-functional, but the distinction between functions within the district is much more strongly expressed than in a neighborhood. Districts typically contain neighborhoods with major public architecture and open spaces, associated priest quarters or civic-ceremonial zones, elite- and commoner-dominated neighborhoods, along with storage complexes, and road/boundary marking features.

16.5 Conclusion

Here we have provided one example of how LiDAR derived high-density point clouds have transformed the spatial understanding of a West Mexican urban center and its linked hinterland. For the ancient city of Angamuco, aerial LiDAR has provided high-resolution data that would have been impossible to achieve with traditional survey techniques in the lab. Through this analysis we have been able to more fully understand the extent, form, and spatial organization/association of ancient architecture at Angamuco along with temporal and functional associations. This new understanding promises to significantly alter interpretations for the development of complex societies in the region and provides a powerful example of the potential of this new technology.

More broadly, as pointed out by Chase et al. (2012), the application of LiDAR techniques parallels the development, application, and impact of radiocarbon dating to archaeology during the 1950s and 1960s and I want to briefly discuss three related aspects of this argument. First, in a historical summary of radiocarbon dating in American Antiquity up to 1985, Taylor points out that a major "thematic" issue in this literature is the acknowledgement that archaeologists were uncomfortable using a methodology that was outside their disciplinary purview. This is a situation similar to the current state of the use of LiDAR in archaeology, given that we are at initialization in terms of the tools and theoretical constructs needed to make use of these new data – and that much of this is coming from other disciplines.

A second related point is that it took almost a decade for radiocarbon dating to gain wide acceptance as a mechanism that allowed archaeologists to move beyond chronology to true theory building, a notion that was intrinsic to the "New Archaeology" (e.g. Binford 1972: 10–11). For LiDAR, we are seeing something similar in that there is a growing recognition that point cloud data yield more then simply placing vegetated areas on an equal footing with those lacking such cover (e.g. the Maya lowlands vs Peru). Instead these three-dimensional landscape pictures are new sandboxes for emerging kinds of sophisticated spatial analysis and modeling.

Finally, in 1967 Glyn Daniel likened the advent of radiocarbon dating to the discovery of "the human species in the nineteenth-Century" (Daniel 1967: 266 as quoted in Taylor 1985: 309). Though he did not pitch it as such, one could clearly argue that Daniel and many others were seeing the wide acceptance of Carbon-14 dating as something akin to a "scientific revolution" in the sense of Thomas Kuhn. Point cloud data allow archaeologists, for the first time, to examine the landscape in the three dimensions exactly as it is experienced by people. If the use and acceptance of radiocarbon dating was a "scientific revolution," LiDAR and similar techniques may very well have a similar impact.

Acknowledgements This work was made possible by a dedicated team of graduate and undergraduate students from North America, Mexico, and Europe. This work was funded by a National Science Foundation award to Christopher Fisher (BCS 0818662), A NASA Space Archaeology award to Stephen Leisz and Christopher Fisher, and Colorado State University. Merrick and Co. provided assistance with LiDAR acquisition and the use of MARS LiDAR software. Trimble provided assistance with GPS and handheld computers. We would like to thank the Instituto Nacional de Antropología e Historia, Mexico, for permits, support, and guidance. Three anonymous reviewers provided useful comments and suggestions.

Appendix 1: LiDAR Technical Data

During the LiDAR data collection, the operator recorded weather conditions, LiDAR operation parameters, and flight line statistics. Before the flight two GPS base stations were established and 4 GPS control points surrounding the area where LiDAR collection took place were established. A five-minute INS initialization was conducted on the ground, with aircraft engines running. Near the end of the mission GPS ambiguities were again resolved by flying within 10 km of the GPS base stations to aid in post-processing of the data. The horizontal datum used was WGS84, the vertical datum was the Gravimetric Geoid (GGMO5). The geoid was converted to ellipsoid heights to orthometric heights. The coordinate system was UTM Zone 14 and all units were in meters. Sixteen ground returns per meter (1 return every 25 cm) were specified in the protocol. Accuracy in the horizontal and vertical were reported at no worse than 0.027m and NSSDA Achievable contour intervals, ASPRS Class 1 achievable contour intervals, and NMAS achievable contour intervals are reported as being 0.05 m. All LiDAR data was processed using MARS proprietary software.

References

Arnauld, M. C., & Faugère-Kalfon, B. (1998). Evolución de la ocupación humana en el Centro-Norte de Michoacán (Proyecto Michoacán, CEMCA) y la emergencia del Estado Tarasco. In V. Darras (Ed.), *Génesis, culturas y espacios en Michoacán* (pp. 13–34). Mexico: Centre d'Etudes Mexicaines et Centraméricaines.

Ashmore, W. A. (1981). Some issues of method and theory in lowland Maya settlement archaeology. In W. Ashmore (Ed.), *Lowland Maya settlement patterns* (pp. 37–69). Albuquerque: University of New Mexico Press.

Beekman, C. (2010). Recent research in Western Mexican archaeology. *Journal of Archaeological Research, 18*, 41–109.

Binford, L. R. (1972). *An archaeological perspective*. New York: Seminar Press.

Cárdenas García, E. (1999). *El Bajío en el Clásico: Análisis Regional y Organización Política*. Zamora: El Colegio de Michoacán.

Chase, A. F., Chase, D. Z., Fisher, C. T., Leisz, S. J., & Weishampel, J. F. (2012). Geospatial revolution and remote sensing LiDAR in Mesoamerica archaeology. *Proceedings of the National Academy of Sciences, 109*(32), 12916–12921.

Chase, A. F., Chase, D. Z., Weishampel, J. F., Drake, J. B., Shrestha, R. L., Slatton, K. C., Awe, J. J., & Carter, W. E. (2011). Airborne LiDAR, archaeology, and the ancient Maya landscape at Caracol, Belize. *Journal of Archaeological Science, 38*(2), 387–398.

Cowgill, G. L. (2004). Origins and development of urbanism: Archaeological perspectives. *Annual Review of Anthropology, 33*, 525–549.

Daniel, G. (1967) *The Origins and Growth of Archaeology*. Crowell, New York.

Darras, V., & Faugère, B. (2005). Cronología de la cultura Chupícuaro. Estudio del sitio La Tronera, Puruagüita, Guanajuato. In E. Williams, P. C. Weigand, L. L. Mestas, & D. Grove (Eds.), *El Antiguo Occidente de México. Nuevas Perspectivas Sobre el Pasado Prehispánico* (pp. 255–281). Mexico: El Colegio de Michoacán-Instituto Nacional de Antropología e Historia.

Feinman, G. (1991). Demography, surplus, and inequality: Early political formations in highland Mesoamerica. In T. Earle (Ed.), *Chiefdoms: Power, economy, and ideology* (pp. 229–262). Cambridge: Cambridge University Press.

Fisher, C. T. (2005). Demographic and landscape change in the lake patzcuaro basin, Mexico: Abandoning the garden. *American Anthropologist, 107*(1), 87–95.

Fisher, C. T. (2007). Agricultural intensification in the lake pátzcuaro basin: Landesque capital as statecraft. In T. Thurston & C. T. Fisher (Eds.), *Seeking a richer harvest: The archaeology of subsistence intensification, innovation, and change* (Studies in human ecology and adaptation, Vol. 3, pp. 91–106). New York: Springer.

Fisher, C. T. (2009a). Abandoning the garden: The population/land degradation fallacy as applied to the lake pátzcuaro basin in Mexico. In C. T. Fisher, J. B. Hill, & G. M. Feinman (Eds.), *The archaeology of environmental change* (pp. 210–231). Tucson: University of Arizona Press.

Fisher, C. T. (2009b). *Technico Parcial: Legados de la resiliencia: La Cuenca de Pátzcuaro Proyecto Arqueológico (Proyecto LORE LPB) 2009* (p. 75). Mexico: Instituto Nacional de Antropología y Historia.

Fisher, C. T. (2010). *Technico Parcial: Legados de la resiliencia: La Cuenca de Pátzcuaro Proyecto Arqueológico (Proyecto LORE LPB) 2010* (p. 100). Mexico: Instituto Nacional de Antropología y Historia, Mexico.

Fisher, C. T. (2011). *Technico Parcial: Legados de la resiliencia: La Cuenca de Pátzcuaro Proyecto Arqueológico (Proyecto LORE LPB) 2011* (p. 110). Mexico: Instituto Nacional de Antropología y Historia.

Fisher, C. T., Hill, J. B., & Feinman, G. M. (2009). Introduction: Environmental studies for twenty-first century conservation. In C. T. Fisher, J. B. Hill, & G. M. Feinman (Eds.), *The archaeology of environmental change* (pp. 1–12). Tucson: University of Arizona Press.

Fisher, C. T., Leisz, S., & Outlaw, G. (2011). LiDAR: A valuable tool uncovers an ancient city in Mexico. *Photogrammetric Engineering and Remote Sensing, 77*(10), 963–967.

Fisher, C. T., Pollard, H. P., Israde-Alcántara, I., Garduño-Monroy, V. H., & Banerjee, S. (2003). A reexamination of human-induced environmental change within the lake patzcuaro basin, michoacan, Mexico. *Proceedings of the National Academy of Sciences, 100*(8), 4957–4962.

Hirth, K. G. (2000). *Ancient urbanism at xochicalco. The evolution and organization of a prehispanic society* (Vol. 1 and 2). Salt Lake City: University of Utah Press.

Joyce, A. (2009). Theorizing Mesoamerican urbanism. *Ancient Mesoamerica, 20*, 189–196.

Marcus, J., & Sabloff, J. A. (Eds.). (2008). *The ancient city: New perspectives on urbanism in the old and new world*. Santa Fe: SAR Press.

Michelet, D. (1996). El origen del reino tarasco protohistórico. *Arqueología Mexicana, 19*, 24–27.

Michelet, D. (2000). Yacatas y otras estructuras ceremonials tarascas en el Malpaís de Zacapu, Michoacán. In J. Litvak & L. Mirambell (Eds.), *Arqueología, historia y antropología. In memoriam José Luis Lorenzo Bautista* (pp. 117–137). México: INAH (Colección -Cientffica 415).

Michelet, D. (2008). Vivir Diferentement. Los sitios de la fase Milpillas (1250–1450 d.C.) en el malpaís de Zacapu (Michoacán). In A. G. Mastache, R. H. Cobean, A. Garcia Cook, & K. G. Hirth (Eds.), *El Urbanismo en Mesoamérica/Urbanism in Mesoamerica vol. 2. Instituto Nacional de Antropología e Historia* (pp. 447–499). University Park: Pennsylvania State University.

Pollard, H. P. (1993). *Taríacuri's legacy*. Norman: University of Oklahoma Press.

Pollard, H. P. (2008). A model of the emergence of the tarascan state. *Ancient Mesoamerica, 19*, 217–230.

Sabloff, J. A. (2008). *Archaeology matters*. Walnut Creek: Left Coast Press.

Sabloff, J. A., & Ashmore, W. (2001). An aspect of Archaeology's recent past and its relevance in the new millennium. In G. M. Feinman & T. D. Price (Eds.), *Archaeology at the millennium: A sourcebook* (pp. 11–37). New York: Kluwer.

Smith, M. T. (2003). *The political landscape: Constellations of authority in early complex polities*. Berkeley: University of California Press.

Smith, M. E. (2006). How do archaeologists compare early states? *Reviews in Anthropology, 35*, 5–35.

Smith, M. E. (2010a). The archaeological study of neighborhoods and districts in ancient cities. *Journal of Anthropological Archaeology, 29*(2), 137–154.

Smith, M. E. (2010b). Sprawl, squatters and sustainable cities: Can archaeological data shed light on modern urban issues? *Cambridge Archaeological Journal, 20*(2), 229–253.

Stark, B. L. (1999). Formal architectural complexes in South-Central Veracruz, Mexico: A capital zone? *Journal of Field Archaeology, 26*, 197–225.

Stark, B. L., & Arnold, P. J., III (Eds.). (1997). *Olmec to aztec: Settlement patterns in the ancient Gulf lowlands*. Tucson: University of Arizona Press.

Taylor, R. E. (1985). The beginnings of radiocarbon dating in American Antiquity: A historical perspective. *American Antiquity, 50*(2), 309–325.

Part V
Archaeological Site Detection and Modeling

Chapter 17
Methods, Concepts and Challenges in Archaeological Site Detection and Modeling

Michael J. Harrower

Abstract This chapter briefly reviews methodological and conceptual issues and debates central to advancement of archaeological site detection and modeling. A wide variety of different survey, sampling and site designation methods have long shaped results of archaeological explorations. Methodological challenges and potential means to better address sociocultural considerations within geospatial prediction/detection modeling are mentioned, and less widely recognized methodological impediments including the Modifiable Areal Unit Problem (MAUP) and the need to address detection errors of omission and commission are discussed.

Keywords Satellite remote sensing • GIS • GPS • Geospatial • Archaeological survey • Landscape archaeology • Regional analysis • Full coverage survey • Predictive modeling • Modifiable Areal Unit Problem

17.1 Introduction

Geospatial technologies have long held promise for detecting and modeling the spatial distribution of archaeological sites and examining individual sites themselves. With the increasingly wide implementation of GPS, GIS, air and spaceborne imagery and technologies, archaeology has arguably entered a new era requiring not only technical expertise but just as importantly critical and reflexive consideration of applications and ramifications. By the dawn of the twentieth century, archaeologists had recognized the utility of air photographs and had begun, for example, to map ancient cities and irrigation networks (Beazeley 1919; Stein 1919). Hopeful expectations following the availability of Landsat imagery, with recognition that archaeology (beyond characterizing environments) would require far higher spatial resolution imagery (Allan and Richards 1983; Ebert 1984), have now begun to be realized. While many obstacles remain, substantial progress has been made and challenges facing rapid and accurate site detection and modeling are now far better understood.

D.C. Comer and M.J. Harrower, *Mapping Archaeological Landscapes from Space*, SpringerBriefs in Archaeology, DOI 10.1007/978-1-4614-6074-9_17, © Springer Science+Business Media New York 2013

As the spatial and spectral resolution of imagery and the sophistication of processing software and algorithm development tools continue to improve, so too will archaeologists' ability to distinguish the world's nearly innumerable vestiges of ancient human activity, including agricultural field systems, settlements and architecture, tombs and monuments small and large. Indeed, rather than predictive modeling of archaeological sites' *probable* locations, direct (semi)automated detection of archaeological sites and features themselves is increasingly possible (see chapters that follow). Yet in addition to the technical challenges to be solved by improved and customized hardware and software, central conceptual/methodological issues and disagreements remain, including:

- What is meant in any particular region by the notion of an archaeological "site."
- Differences regarding the perceived value of opportunistic versus systematic survey.
- Differences regarding the appropriateness of regional sampling-based versus full or 100% coverage survey.
- Failure to acknowledge, address and report detection errors of omission and commission.
- Malleability of results created by the Modifiable Areal Unit Problem (MAUP).
- Need to better integrate social and cultural dimensions of ancient life in geospatial analysis.

17.2   Definitions and Taxonomies of Archaeological Sites

A significant dialogue centered in the 1980s considers challenges in defining what should be deemed an archaeological site, with some advocating site-less archaeological survey and distributional archaeology (Dunnell 1992; Dunnell and Dancey 1983; Ebert 1992). In many areas of the world, landscapes nearly everywhere contain some detectable trace of past human activity such as scatters of lithics and ceramics, campsites and settlements, rockart, and burials of greater or lesser densities. For different applications, archaeologists with widely differing research and cultural resource management objectives are often behooved to define some higher density concentrations of ancient remains as sites while other (perhaps simply less appreciated locations) elude an official designation.

While discarding the designation "site," adopting some other fixed terminological designation, or attempting to simply cover all low and high density areas have yet to offer widely agreed upon solutions (see Terrenato 2004), the wide popularity of landscape-based approaches among archaeologists of widely differing theoretical orientations (e.g., Anschuetz et al. 2001; Ashmore and Knapp 1999; Rossignol and Wandsnider 1992; Ucko and Layton 1999; Wilkinson 2004) illustrates the importance of flexibility with regard to "site" designations and terminologies. Indeed, with the near ubiquitous use of GPS, archaeologists are now able to pinpoint (even piece-plot) a multitude of different types of remains, making decisions about what to record and what not to record (and how) evermore significant. In any particular instance taxonomic designations of differing concentrations of archaeological remains shape research outcomes and necessarily require locally tailored solutions with important ramifications for site detection and modeling.

17.3 Sampling Based Versus Full Coverage Survey

In tandem with arguments for site-less archaeological survey, the late 1970s and 1980s also witnessed the beginnings of two related yet somewhat incongruous approaches, full coverage survey and regional analysis (Fish and Kowalewski 1990; Johnson 1977), for which space technologies now increasingly intercede. While laudable in theory and practicable independently, the goal of full 100% coverage survey and the goal of regional analysis (for areas hundreds or thousands, rather than tens, of square kilometers) exemplifies a central dichotomy between intensive versus extensive survey approaches (Terrenato 2004).

The vaguely defined "full-coverage regional" survey unwaveringly advocated by Kowalewski (2008) may be practicable in cases of easily navigable terrain with limited vegetation where very large teams have many years or decades available, yet full regional coverage is impractical or next to impossible in steep and treacherous, heavily vegetated, or otherwise inaccessible terrain of many regions worldwide. The dense and nearly impenetrable tropical vegetation dominant across parts of the Yucatan (Chase this volume), the steep and dissected mountainous highlands of northern Ethiopia (D'Andrea et al. 2008), and the alternately glaciated and heavily vegetated mountains of southwestern British Columbia (Carlson 1979; Oliver 2007) exemplify only a few the areas in which archaeologists face difficult choices between intensive versus extensive survey often shaped by limited accessibility, time and resources.

While extensive regional approaches are required to construct understanding of the often vast areas covered by various societies from nomads to ancient states, they frequently neglect the rich and more comprehensive understanding of place that intensive full coverage of smaller areas can provide. Indeed, the following section of this volume in many respects exemplifies the various local, regional and intermediate scales within which space-based technologies now assist and intervene, from Menze and Ur's ground-breaking regional detection of ancient northern Mesopotamian settlements (Chap. 18), Tilton, Comer and Chen et al.'s pioneering intermediate-scale detection efforts in California (Chaps. 20 and 21), to the intensive local innovation of Branting's kinematic survey of Kerkenes Da (Chap. 19).

Importantly, statistical sampling strategies are designed to obviate the need to examine every locale, every site, and every house etcetera so that accurate conclusions can be drawn from samples. Yet exclusive conformity to systematic sampling can exhaust considerable time and resources navigating to and searching randomly selected areas while neglecting the utility of opportunistic approaches that draw on the experience of archaeologists, serendipity, and knowledge of local residents who

frequently know and can point surveyors to sites or locations of extraordinary interest. Conversely, as Kvamme (1985) has cogently emphasized, understanding where
sites are requires understanding of where they are not, so at least some systematic
coverage is required to balance opportunistic wanderings if one seeks to gain representative understanding of an area.

17.4   Detection Errors of Omission and Commission

In addition to the aforementioned taxonomic and sampling issues that will retain
central importance as archaeological field data increasingly serve as baseline
training sets for site detection and modeling, errors of omission and commission
and *a priori* probabilities are of similarly central significance. Particularly since
modeling site locations or detecting sites in air or spaceborne imagery is not a
simple exercise, detailed reporting of a particular method's success or lack
thereof is crucial. To view or process a satellite image and identify 100 candidate
locations that appear to be archaeological sites, visit them in the field, and discover some manner of archaeological remains at 85 or more of them belies the
that fact in some areas the probability of encountering archaeological remains is
very high and if one to were to visit 100 randomly selected locations one might
similarly discover nearly 85 sites (Kvamme 2006: 26–31). Concordantly, a predictive model that includes 75% of known sites and 75% of a study landscape has
no predictive value (Kvamme 2006: 28). Either data for non-site locations or data
on background values of potentially predictive variables is therefore required to
properly evaluate the efficacy of any variable or method of site prediction or
detection. Moreover, reporting errors of commission, in which detection procedures identify a location as an archaeological site when there is no actual site
present (false positive), and errors of omission, in which a known site is not
detected or included (false negative) is critical to evaluating results. Many direct
detection efforts are highly promising yet still at a relatively incipient stage, and
both types of errors will eventually need to be fully evaluated and reported to
properly assess the level of success, and facilitate necessary adjustments.

17.5   The Modifiable Areal Unit Problem

The Modifiable Areal Unit Problem (MAUP) is of similarly crucial importance for
modeling and detection. As Openshaw (1981, 1983) described, MAUP describes a
type of category of capricious results that arise because of arbitrary or discretionary
scale, boundary or data categorization choices (Dungan et al. 2002; Fotheringham
and Wong 1991). Indeed, archaeologists frequently face scale and data categorization
alternatives that can dramatically impact analytical results potentially contributing

to what can memorably be deemed archaeological gerrymandering (Harrower 2009). From the size and shape of study area and related sampling units, the number and type of landform, landcover, or soil classes, to (if categorized) the number of classes and intervals used for variables such as slope and aspect, archaeologists continually face options and choices with important potential impacts on results. As Kvamme (2006: 7, 21) cogently noted, "blue-line" water features on topographic maps are frequently the result of arbitrary/discretionary decisions (often traditionally based on somewhat subjective interpretation of air photos). Similarly, Strahler or Shreve stream order representations of drainage networks critically depend on where one decides to begin (or what one decides qualifies as) first order streams.

Not only should categories be defined as objectively and independently as possible from the archaeological data at hand, (such as soil types, landform, or landcover categories defined by explicit pedological, geomorphological, or ecological criteria) but analysts should also pay careful attention to how altering the number of intervals and boundaries of categories impacts the outcomes of analyses. Importantly, variables for which continuous data is available, such as slope and aspect, are best analyzed using their full richness as continuous variables rather than classifying them into arbitrary/discretionary categories which can add malleability and subjectivity to results.

17.6 Social and Cultural Dimensions of Ancient Life

Geospatial and predictive modeling analyses have often been subject to critique for insufficiently including social and cultural considerations. While there can be no clear subdivision between nature and culture, environmental variables are often more readily quantified and included in geospatial analyses in comparison with more abstract and often more qualitative (yet just as important) social, cultural and political considerations. Geospatial analyses cannot, of course, be taken as standalone representations of ancient societies and cultures. Nevertheless, geospatial technologies do serve an extraordinary useful purpose and continual efforts to better integrate philological, epigraphic, art historical, lithic, ceramic, archaeobotanical, zooarchaeological and a virtually innumerable plethora of other sources of information with geospatial research holds promise to better address social dimensions of ancient life.

Indeed, since datasets for material remains sites contain are often challenging and time consuming to produce, archaeological sites are sometimes analyzed as simple point entities across landscapes with little integration of their true richness as ancient places. This need not, and undoubtedly increasingly will not, be the case as geospatial technologies become better integrated within standard archaeological practice. Oftentimes the most informative results arise when predictive, locational or other models fail to accurately represent ancient choices prompting analysts to identify and consider new dimensions of ancient life that are insufficiently addressed.

218   M.J. Harrower

References

Allan, J.A., and T.S. Richards (1983) Use of Satellite Imagery in Archaeological Surveys. *Libyan Studies* 14:4–8.

Anschuetz, K. F., Wilshusen, R. H., & Scheick, R. H. (2001). An archaeology of landscapes: Perspectives and directions. *Journal of Archaeological Research, 9*(2), 157–211.

Ashmore, W., & Knapp, B. (Eds.). (1999). *Archaeologies of landscape: Contemporary perspectives.* Oxford: Blackwell.

Beazeley, G. A. (1919). Air photography in archaeology. *The Geographical Journal, 53*(5), 330–335.

Carlson, R. L. (1979). The early period on the Central Coast of British Columbia. *Canadian Journal of Archaeology, 3*, 211–228.

D'Andrea, A. C., et al. (2008). The Pre-Aksumite and Aksumite settlement of northeastern Tigrai, Ethiopia. *Journal of Field Archaeology, 33*(2), 151–176.

Dungan, J. L., et al. (2002). A balanced view of scale in spatial statistical analysis. *Ecography, 25*, 626–640.

Dunnell, R. C. (1992). The notion site. In J. Rossignol & L. Wandsnider (Eds.), *Space, time, and archaeological landscapes* (pp. 21–37). New York: Plenum.

Dunnell, R. C., & Dancey, W. S. (1983). The siteless survey: A regional scale data collection strategy. *Advances in Archaeological Method and Theory, 6*, 267–287.

Ebert, J. I. (1984). Remote sensing applications in archaeology. *Advances in Archaeological Method and Theory, 7*, 293–353.

Ebert, J. I. (1992). *Distributional archaeology.* Albuquerque: University of New Mexico Press.

Fish, S. K., & Kowalewski, S. (Eds.). (1990). *The archaeology of regions: A case for full-coverage survey.* Washington, DC: Smithsonian Institution Press.

Fotheringham, A. S., & Wong, D. W. S. (1991). The modifiable areal unit problem in multivariate statistical analysis. *Environment and Planning A, 23*, 1025–1044.

Harrower, M. (2009). *Archaeological gerrymandering: The power and pitfalls of GIS* (pp. 40–41). Backdirt: Annual Review of the Cotsen Institute of Archaeology at UCLA.

Johnson, G. A. (1977). Aspects of regional analysis in archaeology. *Annual Review of Anthropology, 6*, 479–508.

Kowalewski, S. (2008). Regional settlement pattern studies. *Journal of Archaeological Research, 16*, 225–285.

Kvamme, K. (1985). Determining empirical relationships between the natural environment and prehistoric site locations. In C. Carr (Ed.), *For concordance in archaeological analysis* (pp. 208–239). Arkansas: Westport Publishers.

Kvamme, K. (2006). There and back again: Revisiting archaeological locational modeling. In M. W. Mehrer & K. L. Wescott (Eds.), *GIS and archaeological site location modeling* (pp. 4–34). Boca Raton: CRC Press.

Oliver, J. (2007). Beyond the water's edge: Towards a social archaeology of landscape on the Northwest Coast. *Canadian Journal of Archaeology, 31*, 1–27.

Openshaw, S. (1981). The modifiable areal unit problem. In N. Wrigley & R. J. Bennett (Eds.), *Quantitative geography: A British view* (pp. 60–69). London: Routledge and Kegan Paul.

Openshaw, S. (1983). *The modifiable areal unit problem.* Norwich: Geo Books.

Rossignol, J., & Wandsnider, L. (Eds.). (1992). *Space, time, and archaeological landscapes.* New York: Plenum.

Stein, A. (1919). Air photography of ancient sites. *The Geographical Journal, 54*, 200.

Terrenato, N. (2004). Sample size matters! The paradox of global trends and local surveys. In S. E. Alcock & J. F. Cherry (Eds.), *Side-by-side survey: Comparative regional studies in the Mediterranean World* (pp. 36–48). Oxford: Oxbow Books.

Ucko, P. J., & Layton, R. (Eds.). (1999). *The archaeology and anthropology of landscape: Shaping your landscape.* London: Routledge.

Wilkinson, T. J. (2004). The archaeology of landscape. In J. Bintliff (Ed.), *A companion to archaeology* (pp. 334–357). Oxford: Blackwell.

Chapter 18
Multi-Temporal Classification of Multi-Spectral Images for Settlement Survey in Northeastern Syria

Bjoern H. Menze and Jason A. Ur

Abstract Nearly all Near Eastern surveys have employed some form of satellite remote sensing, although usually at coarse resolution or for very limited study areas only, and nearly always in a qualitative fashion. In this chapter we briefly review the multi- and hyper-spectral remote sensing approaches that have been used in archaeological survey in the Near East, and report on a novel satellite remote sensing approach we recently developed. This approach recognizes anthropogenic sediments via scenes from multi-spectral ASTER images using a multi-temporal classification strategy, guided from results of a visual inspection of CORONA images. We apply it to the Khabur Basin, in north-eastern Syria, returning a probabilistic map of anthrosols that is indicating the locations of some 10,000 settlement sites of all times – from the eighth millennium B.C. to modern – at a resolution of 15 m for an area of about 22,000 km^2 and with an accuracy that is comparable to modern ground survey. This makes it, to the best of our knowledge, the largest systematic satellite imagery based survey in archaeology. Our multi-temporal classification strategy can integrate information from any other multi- or hyper-spectral sensor, and it will easily generalize to other related detection tasks in archaeological remote sensing.

Keywords Multispectral classification • Multi-temporal classification • Settlement size • Settlement pattern

18.1 Patterns of Sites and Soils

Assessment of the scale and spatial distribution of human communities in past societies has been a key objective for archaeological research over the last 50 years, especially in the Near East, where questions of the origins of urbanism, the state, and empires are almost always approached via a regional perspective (Wilkinson 2003), and most commonly using the methods of archaeological survey. The pioneering

D.C. Comer and M.J. Harrower, *Mapping Archaeological Landscapes from Space*, SpringerBriefs in Archaeology, DOI 10.1007/978-1-4614-6074-9_18, © Springer Science+Business Media New York 2013

surveys of the twentieth century focused on the top of the settlement hierarchy, in the form of the largest mounds, but it is now appreciated that by overlooking smaller sites, such an approach can produce misleading portraits of settlement systems, particularly those of non-urbanized phases (Wilkinson et al. 2004). To map these smaller sites, and to pursue survey on the ground more efficiently, nearly all Near Eastern surveys have employed some form of satellite remote sensing, although usually at a resolution too coarse for resolving detailed archaeological features (Menze et al. 2006) or for very limited study areas (Altaweel 2005; Wilkinson et al. 2006) only, and nearly always in a qualitative fashion. In the following, we will describe a remote sensing approach that recognizes anthropogenic sediments, which makes it an ideal tool for mapping settlement sites from all levels of the settlement hierarchy.

Anthropogenic soils are a characteristic indicator of long-lasting settlement activity in many Near Eastern landscapes (Wilkinson 2003). In the alluvial and largely treeless plains of the northern arc of the Fertile Crescent – as in many other landscapes of the Near East – the primary building material was mud brick. As dwellings became dilapidated, their walls were partially dismantled and flattened, and new structures were built atop their remains. In this manner, settlements grew vertically in the centuries or millennia that they were occupied, and soils at the site transformed to characteristic anthrosols. The largest mounds – known as "tells" in Arabic – still rise to heights of dozens of meters, and represent well known landmarks in many regions of the Near East. The presence of anthropogenic soils is a defining feature of these large mounds, but also many smaller settlement sites. During millennia of human occupation these anthrosols formed from organic waste and the decay of mud-brick architecture. Their texture and hydrological and reflective properties often differ significantly from the surrounding soils, because they have a higher amount of organic components, a finer texture and a lighter appearance (Galiatsatos 2004). This makes them visible not only on the ground in regional surveys, but also in satellite imagery.

18.2 Archaeological Survey and Remote Sensing in the Near East

In particular, imagery from the declassified CORONA reconnaissance missions has proven useful for detecting sites and landscape features (Fowler 2004; Ur this volume; Casana and Cothren, this volume). These grayscale panchromatic photographs were acquired in the visible range and provide spectacular views of northern Mesopotamian landscapes of the 1960s and 1970s prior to disturbance by modern development and agricultural expansion (Ur 2003). Imagery from the CORONA program is extensive but still somewhat limited in spatial and temporal extent; for many areas of the Near East, scenes from only one or two missions are available, and many of them may have been acquired under conditions that are sub-optimal for archaeological site visibility.

This limitation in the availability of appropriate satellite imagery can be overcome in part by newer sensors that revisit areas frequently throughout the year, as the archaeological matrix of anthrosols is also visible in the infrared spectrum recorded in multi-spectral satellite images (Altaweel 2005). Unfortunately, due to the high specificity of the multi-channel signal, the signatures of the structures of interest and their surroundings change to a much larger extent during the year than the image intensities in standard monochrome imagery. At the same time, characteristic patterns in the multi-spectral signatures are difficult to infer for human observers due to the high dimensionality of the data and a relatively low spatial resolution of the images. As a consequence, the interpretation of imagery from cameras such as Landsat or ASTER has only complemented the visual inspection of high-resolution aerial imagery. So far, the analysis of multi-spectral images has not had a significant impact on regional surveys conducted in the landscapes of the Near East, although their widespread use has been anticipated since data from the first multi-spectral sensors was available several decades ago.

Recently, we presented an approach that is able to overcome some of these limitations (Menze and Ur 2012), testing it in a region in northeastern Syria (Fig. 18.1). It makes explicit use of the large satellite image databases that have been acquired with various multi-spectral sensors over the past decades, using a novel multimodal classification strategy for multi-spectral images and requiring only limited ground control. As site visibility is closely connected to ground conditions, it may be insufficient to rely on a single satellite image for a given area. In ASTER images, for example, the spectral signature of the anthrosols changes inter-annually and throughout the year, related to rainfall fluctuations and the agricultural cycle (Fig. 18.2, left column). This variability is a significant problem when training a machine learning algorithm that is using the spectral signal of a few known archaeological sites as input to predict the location of other settlements in the same region; areas with similar hydrological conditions or those that are partially covered by the same vegetation as the archaeological sites may show up as false positives, and the resulting probability maps will always have a large amount of random "noise" (Fig. 18.2, right column).

For this reason, our multi-temporal approach uses the information from all multi-spectral images covering a region of interest. It only highlights those locations that *consistently* – over long observational periods – show the same spectral signature as nearby archaeological sites. We will provide details of this approach in the following, and describe its use for mapping sites in the upper Khabur Basin at a large scale. Aspects where further information can be found in the original publication are indicated by a reference to Menze and Ur (2012).

18.3   Mapping Settlement Sites in the Upper Khabur Basin

The upper Khabur basin, a region in northeastern Syria, is known as a "landscape of tells" (Wilkinson 2003). It extends about 200 km in east–west direction and 100 km from north to south (Fig. 18.1). This basin is a critical locus for the study of the origins and development of urbanism and the Near East, and the organization of settlements under early states and empires in northern Mesopotamia.

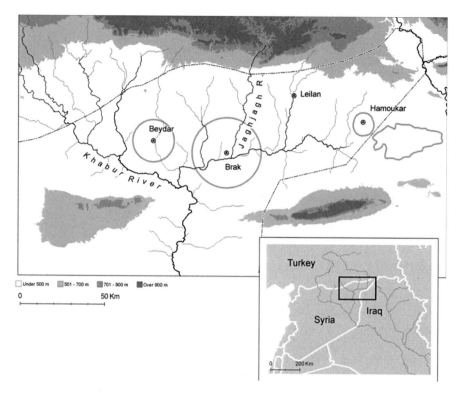

Fig. 18.1 Localization. The Upper Khabur Basin with rainfall isohyets and major archaeological surveys indicated; from west to east: Tell Beydar Survey, Tell Brak Survey, Tell Hamoukar Survey, North Jazira Project. Inset: Localization in Hassake Province, northeastern Syria

18.3.1 A Multi-Temporal Classification Strategy

To develop the classifier, we begin with 1680 archaeological sites identified on CORONA images all throughout our search area; about 10% of them had previously been verified by ground survey (Ur 2010). The image dataset comprises 160 multi-spectral ASTER 1B images that have been processed to sensor units, acquired between 2003 and 2007 and spread evenly across all seasons.

Some parts of the survey area are covered by more than ASTER 40 scenes, while others have as few as four. Clouded areas were removed from the analysis, and all ASTER image were spatially registered to the reference image (composed from 10 m resolution SPOT images) that had previously been used for the CORONA images. ASTER images have 14 spectral channels (Fig. 18.2) with reflectances from visible red (V) and near-and short-wave infrared (NIR, SWIR) to emissions in long-wave and thermal infrared (TIR); the spatial resolution differs from 15 (V) to 90 m (TIR). The detection problem is transformed into a binary classification task by amending the known archaeological sites with a second group of randomly chosen

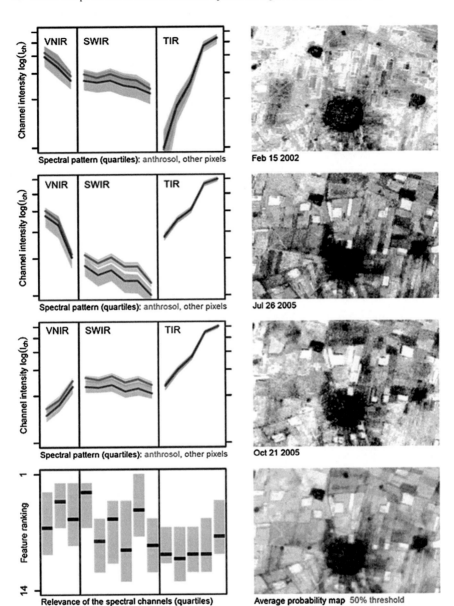

Fig. 18.2 The multi-temporal classification approach. Seasonal variation in spectral patterns in ASTER images and spatial probability maps for the Hamoukar region. Spectra for anthrosols (*red*) and other pixels (*blue*) are shown in the *left column*; *lines* represent medians, *boxes* quartiles (TIR has a different offset than the other channels). The anthrosol probability are shown in the *right column* (*black* 100%, *white* 0%; 50% threshold outlined *red*). The figure in the *bottom left* shows what spectral channels were relevant for the classifier (*boxes* represent quartiles; *black lines* median) (Menze and Ur 2012). The final average probability map (*bottom right*) is obtained by averaging five probability maps (including the three shown here)

locations that are assumed to be non-archaeological sites. Through this random sampling, a few pixels from archaeological sites may be assigned erroneously to the "non-settlement" class. The classifier we choose, however, is able to deal with a low number of such training samples with false labels. For every pixel at the locations of both the "settlement" and "non-settlement" class, a number of descriptors of the spectral information are recorded: the original reflectances from the 14 ASTER channels, three standard vegetation indices representing differently normalized reflectances, and the correlation of the full spectrum with different template spectra from a spectral library representing soil types, water, minerals, and vegetation.

Using these descriptors for "settlement" and "non-settlement" locations within the field of view of an ASTER image, we train a "random forest" classification algorithm to discriminate between the two classes. This nonlinear random forest classifier requires little parameter optimization during training, and we can adapt the algorithm individually to every satellite image available for our study region (Menze and Ur 2012). This adaptive retraining of the classifiers makes prediction more robust against changes of the spectral signal due to conditions of the surface or atmosphere. When presented with the spectral descriptors of other pixels from the same image in a spatially blocked cross-validation, the classifier returns a probability indicating how likely anthrosols of the "settlement" class are present at those locations. For every ASTER image we finally obtain a probability map with continuous values which are close to zero for "non-settlement" pixels and close to one for anthrosols and "settlement" sites. In spite of the computationally expensive cross-validation, processing a full ASTER image using the random forest classifier requires no more than 2–3 h on a standard personal computer.

After all ASTER images are classified, the resulting probability maps are fused (Fig. 18.2, right column). This produces an average signal that is largely free of noise from short-term variations of the spectral signal and that highlights only those locations that consistently show the same spectral signature as nearby archaeological sites in the "settlement" class, reducing the effect of short-term variation observed when classifying single images. As the average probability still fluctuates somewhat based on soil, soil cover, and how those changes are represented in the local training data, the probability map is represented as a gray scale map and inspected visually in a manner similar to, for example, a visual inspection of a CORONA image (Fig. 18.3).

18.3.2   Comparison with Archaeological Survey and Validation

The visual inspection of the map returns a total of 14,312 sites[1] that stand out from the spectral signature of their surrounding and, typically, show the round morphology of early settlement sites.

---

[1] Anthrosol maps and all other data products are available for download from http://hdl.handle.net/1902.1/17731

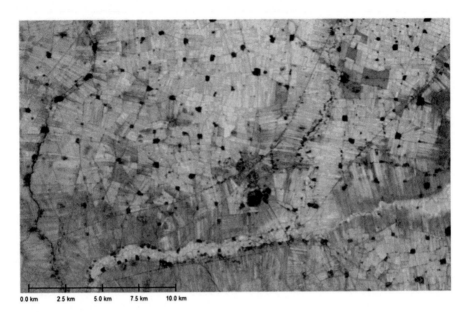

Fig. 18.3 Anthrosol probability map. Anthrosol probability map for the Tell Brak region in the central area of the basin. *Dark values* indicate high probabilities, *bright* values low. Tell Brak is visible as the *dark* area in the south-central part of the map

An automated routine determines the extensions of every recorded site and measures its area in the anthrosol map, and its elevation and volume in a digital elevation model (Menze and Ur 2012). In this analysis, about 12,000 sites are larger than 1 ha, and a total area of 856 km² displays signs of ancient or present settlement. This amounts to about 4% of the total search area of 22,000 km². Out of all sites, 9,529 are significantly mounded and about 5,350 (or 461 km²) show further signs of anthrosols or early settlements when examined in high resolution SPOT or CORONA images.

To further validate results, we compared them against data of several archaeological surveys in the region (Fig. 18.1) and against other proxies of long-term settlement activity with ground truth available at the pixel level.

When comparing results with ground truth from large regional surveys, about 90–94% of the 750 sites recorded around Tell Brak, Leilan, Hamoukar and Beydar can be found in the anthrosol map (Menze and Ur 2012). In a test in the Tell Hamoukar and Tell Beydar survey areas, about one third of the sites identified in the probability map could not be associated with sites previously mapped on the ground (Menze and Ur 2012). In most cases, however, evaluating these additional sites without ground control remains difficult as modern settlement often show a spectral signature similar to the one of early settlements in their surroundings. Many "false positive" represent modern villages that can be recognized through visual inspection of standard high-resolution imagery, and the measurement of site volume via the SRTM terrain dataset (for details about the extended analysis of these data, see Menze and Ur 2012).

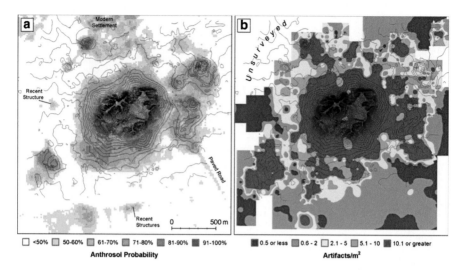

Fig. 18.4 Tell Brak validation. Anthrosol probability map at Tell Brak ((**a**); compare with Fig. 18.3) and sherd density recorded in the Tell Brak Suburban Survey (**b**) (Ur et al. 2007). Outlines found on the ground and anthrosols found via multispectral classification are in excellent agreement

In the case of Tell Brak, the major settlement mound of the Basin, we can compare the anthrosol probability with ground truth that is available nearly at pixel level (Ur et al. 2007). We find areas of high surface artifact density and high probability of anthropogenic soils to correlate almost precisely (Fig. 18.4). These results confirm that the anthrosol map not only identifies archaeological sites, but also that it provides a reliable estimate of sites' shapes and extents, revealing – for example – the border of lower towns around high mounded sites, which otherwise require intensive field surveys.

18.4   Generalization and Limitations

With more than 14,000 potential sites recorded in an area of 22,000 at 15 m resolution, our search for settlements in the Khabur Basis is – to the best of our knowledge – the largest systematic satellite imagery based survey in archaeology. While we focused on anthrosols that relate to Near Eastern settlements, the multitemporal classification approach can be easily generalized for other detection tasks in archaeological remote sensing and for detection task at other scales and relying on satellite images at other spatial resolutions. Moreover, by fusing individually generated probability maps it can integrate information from different multi-spectral sensors, such as Landsat with its long temporal coverage or Hyperion with its high spatial and spectral resolution, in a straightforward fashion. Introducing sub-classes for "settlement" and "non-settlement", or weighting the individual probability maps according to their classification accuracy on the available ground truth may further improve the performance of the presented classification approach.

For generalizing the settlement detection to areas other than the Khabur Basin, some ground control on the location of settlement sites is required. This may stem from a previous analysis of high resolution satellite or aerial images and published information from previous archaeological surveys, or from the analysis of digital elevation models that reveal high mounded sites in many fluvial areas in the Near East (Menze et al. 2006). Combining the multi-spectral analysis with the analysis of a digital elevation models also allows for identifying and removing possible false positive sites, as well as for ranking the sites with respect to volume and their general long-term relevance (Menze and Ur 2012). In general, this short summary article has only presented the basics of this work – the reader is referred to Menze and Ur 2012 for a fuller treatment.[2] Using this approach we envision a nearly comprehensive map of sedentary human settlement for the fluvial plains of Northern Mesopotamia using the proposed approach, with results to be used in the integration of published findings, the planning of new surveys, and heritage management at international scale.

Endnotes and Acknowledgments This text is based on the original manuscript published in PNAS (Menze and Ur 2012). The work has been supported by funding to BHM from the Fritz-Thyssen-Stiftung, and the German Academy of Sciences Leopoldina (LPDS 2009–10). ASTER scenes were provided at no cost through NASA's educational user program.

References

Altaweel, M. (2005). The use of ASTER satellite imagery in archaeological contexts. *Archaeological Prospection, 12*, 151–166.

Fowler, M. J. F. (2004). Archaeology through the keyhole: The serendipity effect of aerial reconnaissance revisited. *Interdisciplinary Science Reviews, 29*, 118–134.

Galiatsatos, N. (2004). *Assessment of the CORONA series of satellite imagery in landscape archaeology: A case study from the Orontes Valley, Syria*. Ph.D. thesis, Durham University, Durham.

Menze, B. H., & Ur, J. A. (2012). Mapping patterns of long-term settlement in the Near East at a large scale. *PNAS, 109*(14), E778–E787.

Menze, B. H., Ur, J. A., & Sherratt, A. G. (2006). Detection of ancient settlement mounds: Archaeological survey based on the SRTM terrain model. *Photogrammetric Engineering and Remote Sensing, 72*(3), 321–327.

Ur, J. A. (2003). CORONA satellite photography and ancient road networks: A northern Mesopotamian case study. *Antiquity, 77*, 102–115.

Ur, J. A. (2010). *Urbanism and cultural landscapes in northeastern Syria: The Tell Hamoukar survey, 1999–2001*. Chicago: Oriental Institute Publications.

---

[2] The publication is freely available from http://www.pnas.org/content/109/14/E778, the preprint from http://nrs.harvard.edu/urn-3:HUL.InstRepos:8523994

Ur, J. A., Karsgaard, P., & Oates, J. (2007). Urban development in the ancient Near East. *Science, 317*(5842), 1188.

Wilkinson, T. J. (2003). *Archaeological landscapes of the Near East*. Tucson: University of Arizona Press.

Wilkinson, T. J., Ur, J. A., & Casana, J. (2004). From nucleation to dispersal: Trends in settlement pattern in the northern Fertile Crescent. In J. Cherry & S. E. Alcock (Eds.), *Side-by-side survey: Comparative regional studies in the Mediterranean World* (pp. 198–205). Oxford: Oxbow Books.

Wilkinson, K. N., Beck, A. R., & Philip, G. (2006). Satellite imagery as a resource in the prospection for archaeological sites in Central Syria. *Geoarchaeology, 21*, 735–750.

Chapter 19
New Geospatial Technologies Leading to New Strategies: The Case of Kerkenes Dağ, Turkey

Scott Branting

Abstract Innovations in geospatial technologies have exploded over the past several decades, significantly impacting archaeological research. At Kerkenes Dağ, a massive late Iron Age city in Central Turkey, a variety of these technologies have been employed over the past two decades. The techniques used include: aerial photography, satellite imaging, magnetometry survey, resistivity survey, and micro-topographic GPS survey. The availability of this data has allowed a map of the ancient city to be developed and new strategies and methods to be devised, including archaeological transportation modeling using both Agent-Based Modeling (ABM) and Transportation Geographic Information Systems (GIS-T), to understand how the city was used by its inhabitants. The potential impacts of these new strategies for both archaeological interpretation and site management are explored in this chapter.

Keywords Kerkenes Dağ • Turkey • Geophysics • Magnetometry • Resistivity • Global Positioning System (GPS) • Aerial photography • Satellite images • Transportation modeling • Transportation Geographic Information Systems (GIS-T) • Agent-Based Modeling (ABM) • Repast Simphony • Simulation • Time-Geography

19.1 Introduction

Changes in technology have long influenced what archaeologists do and how they do it. As a student, I remember sitting in Robert and Linda Braidwood's laboratory in the basement of the Oriental Institute and hearing about how the shift to airplanes from boats, as a means of transportation to and from the field, impacted not only on logistics but also recording and publication. No longer were weeks of travel built into the end of the field schedule, where notes and plans could be checked aboard ship and drawings drawn and reports written. Instead, the trip back is now measured in hours before returning to the busy schedule of the University.

D.C. Comer and M.J. Harrower, *Mapping Archaeological Landscapes from Space*,
SpringerBriefs in Archaeology, DOI 10.1007/978-1-4614-6074-9_19,
© Springer Science+Business Media New York 2013

Changes in certain types of technologies have had a greater impact on archaeology than changes in other technologies. Changes to technology impacting imaging, measurement, computation, logistics and publication have revolutionized all areas of archaeology including reconnaissance, recording, disseminating results, and monitoring sites and regions. This impact, as the Braidwood's recollections reflect, is not something new to archaeology, despite the explosion of new technological advances in all of these areas over the past few decades.

Changes in geospatial technologies have, for more than a century, had far reaching implications for archaeology. Innovations such as aerial photography were instrumental in changing aspects of archaeology early in the twentieth century (Bewley and R czkowski 2002). Charles Lindbergh, the pioneering aviator, once said while doing aerial surveys that "from my ship I can find one undiscovered ruin for every one that has been located from the ground" (Weyer 1929: 54). When satellite imagery became publically available in the 1970s, it broadened the spatial scope of what aerial photography had accomplished in allowing archaeologists to situate their research within broader landscape perspectives (Adams 1981: 33).

From the last decades of the twentieth century, the pace of geospatial technological innovation has exploded, as seen throughout this volume. High-resolution satellite images now show very small features in a landscape. Formerly classified governmental spy satellite data can show slightly larger features in the landscape from half a century ago. New measuring technologies such as GPS and LIDAR have revolutionized how we locate things within the landscape and allow us to digitally reconstruct that landscape (White this volume). Portable geophysical equipment allows us to see beneath the surface of the ground to locate buildings and structures in order to better plan our research designs. As has been argued elsewhere, the explosion of change within geospatial technologies represents a fundamental shift in how we do archaeology (Branting and Trampier 2007).

The question, then, is what can be done with all this new geospatial technology, in terms of archaeological research or site management strategies, that was previously impossible? Obviously, we have expanded our abilities to more rapidly collect ever-higher resolutions of data. We can better and more precisely locate places to visit during surveys, areas to target during excavations, and identify impacts to both sites and regions as part of a management program. But are there new things we can do? Things that our archaeological predecessors, like the Braidwoods, could perhaps only have dreamed of? The answer is yes, though not every novel idea will prove useful or practical. To illustrate this potential I offer as an example the work we've undertaken using a variety of geospatial technologies at Kerkenes Dağ in central Turkey over the past 20 years.

19.2   Geospatial Technologies at Kerkenes Dağ

Kerkenes Dağ was once a very large walled city 271 ha in size. It was built as a new foundation, likely by the Phrygians, and inhabited for around half a century before being destroyed in the mid-sixth century B.C. After its destruction, an event accompanied by heavy burning across the city, the site of the city was largely abandoned. When coupled with a lack of significant overburden from erosion, this has produced a situation where most of the original floors, streets and surfaces are buried only half a meter or so beneath the surface of the ground. Walls from individual structures as well as the walls enclosing the 757 urban blocks and compounds laid out within the city are evidenced as subtle topographic variations in the modern ground surface or by scatters of stone from their foundations.

Since 1993, many geospatial technologies have been employed at Kerkenes Dağ in order to map out the buried city as both a precursor to excavation and to provide a broader context for situating localized excavations (Summers and Summers 2010). Initially, aerial photography was used. Every square meter of the surface of the site was covered with a visual record at a variety of scales. More recently, these photographs have been augmented with high-resolution satellite data. This aerial perspective provides a unique vantage point from which to make better sense of what might not be intelligible on the ground. What appear to be bits and pieces of walls and differences in vegetation can be brought together by looking at the area from high above.

In addition, over an 8-year span between 1995 and 2002, 90% of the rugged terrain of the city was surveyed by magnetometry. Ongoing resistivity survey since 2002 is further refining our knowledge of the structures in certain areas. These two geophysical techniques provide complimentary ways to look beneath the surface of the ground. The lines of buried walls and structures, which might only protrude above the surface of the ground in bits and pieces or which might be completely buried, can be located and mapped in much greater clarity by using electricity or geomagnetics. In many instances, the use of such techniques can minimize the amount of excavation necessary to understand what lies buried beneath the soil. The extent of these surveys at Kerkenes Dağ was groundbreaking and it provides an important basis for planning out the buildings, blocks, and compounds of this urban landscape (Kvamme 2003; Branting 2004).

Complementing the photographic and geophysical surveys, a detailed GPS micro-topographic survey of the entire site was accomplished over 4 years from 1997 to 2000 (Figs. 19.1 and 19.2). Over one and a half million data readings were collected and used to interpolate a detailed topographic surface accurate to 10–25 cm (Branting and Summers 2002). While collected on the ground using various configurations of survey grade Trimble GPS receivers, the resulting point cloud of data is similar to the results of both terrestrial and airborne LIDAR in many respects. Two major differences are of course the time that it took to collect and that the GPS data points were collected with an understanding of what might be features of archaeological importance. More data points were intentionally collected at critical break points or on and around protruding features from the ancient city, something more easily done while at the feature rather than at a distance away from it. This very detailed model of the surface of the ground allows us to peel back concealing vegetation that may block our view of the ground in the aerial photographs. It can also provide a three dimensional perspective to the geophysical results, allowing changes in slope and elevation to be used to aid in our interpretation of the configurations of buried features.

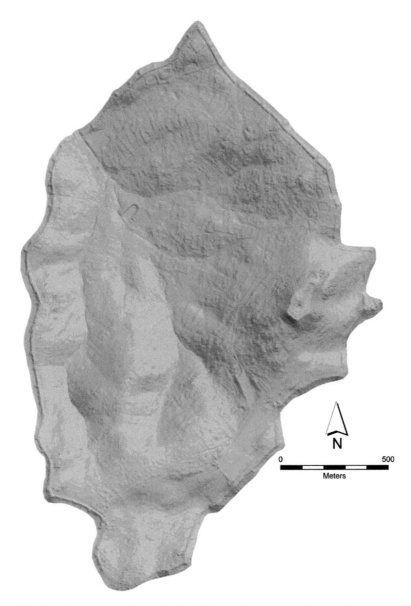

Fig. 19.1 The micro-topographic surface for Kerkenes Dağ generated from a cloud of one and a half million discrete GPS data points. The surface is of comparable accuracy to that of LIDAR (±10–25 cm), though it was collected on foot rather than remotely

Combining the data from these geospatial technologies allowed a detailed plan of the city to be constructed (Branting 2004). Individual buildings and rooms can be identified within and beyond the walled extents of the urban blocks and compounds. Excavation can then target particular portions of structures or areas

Fig. 19.2 An oblique view of the micro-topographic surface showing most of the 271 ha area within the city walls. The remains of the large city wall, the walls of the urban blocks, and even some buildings are clearly visible across the wide extents of the city

to address particular research questions (Summers et al. 2004). However, with the datasets in hand, the question was raised of what more could be done with them? Were there ways that they could be leveraged to do more than just provide a map of the form of the city and its buildings? Could we move past form and attempt to use them to get at the functions of these buildings and activity areas or to the social dynamics at play within the city when it was inhabited? Obviously, excavation is needed to provide information on the artifacts and ecofacts within some of these contexts, essential elements in understanding activities and social dynamics. However, could these datasets leverage that excavated data to provide hypotheses of social function that could subsequently be tested via additional excavation? Could they also be used to help in the ongoing management of the archaeological site?

19.3 New Strategies for Using Geospatial Technologies

One avenue of research utilizing the derived city plan and micro-topographic surface for these purposes proved exceptionally fruitful. Archaeological transportation analysis, developed at Kerkenes Dağ, provides a means by which virtual people can be placed within the digital plan of the city in order to model how the city may have been used (Branting 2004, 2007, 2012). This allows analysis of the buildings and structures as well as the streets and other "empty spaces" from the level of the individual pedestrian (Smith 2008: 218–219; Joyce 2009: 195). Unlike access graphs within a spatial syntax framework (Hillier and Hanson 1984; Hillier 2008), where schematized diagrams of interconnected nodes (places) and lines (connections between them) are used to investigate the interconnectedness of these places, this analysis uses models of the physiology of human pedestrians of different sexes and ages (McDonald 1961; Pandolf et al. 1977; Sun et al. 1996; Santee et al. 2001; Branting 2012; Kantner 2012). At Kerkenes Dağ they are applied to the real-world geographical configuration of the different routes and spaces within the city plan at the scale of the given walker's gait, ca. 30 cm long segments, in order to avoid known issues of scale in such modeling (Kvamme 1990).

The resulting derived costs can be used in either a least-cost or stochastic-based route allocation algorithm to determine optimizing or satisficing routes for people to take based on different scenarios (Branting 2012; White and Surface-Evans 2012). To date this has been accomplished using either agent-based modeling (ABM) through the SHULGI program developed within Repast Simphony or more traditional transportation modeling within a transportation GIS (GIS-T) framework (Branting et al. 2007; Branting 2004, 2012). ABM provides an extremely attractive option for this work, both for allowing individual agents to exhibit independent autonomy in terms of decision making during the simulations and for modeling the interactions and emergent phenomena that arise from those interactions between the agents and the surrounding network (Bonabeau 2002; Branting 2012).

The routes derived from either form of this analysis can be utilized within a time-geography theoretical framework to interpret the movement of individual pedestrians within the simulations as well as aggregated bundles of them forming households, social networks, or institutions (Hägerstrand 1970; Pred 1986, 1990; Branting 2004, 2010). Figure 19.3 shows a result of archaeological transportation analysis at Kerkenes Dağ, where the thickness of the street lines is reflective of the number of virtual pedestrians that walked along that street during a particular simulation. Similar techniques have since been applied to studies on a regional level as well (White and Barber 2012). Figure 19.4 shows a different way to view the results of the analysis. Here the numbers of virtual pedestrians that pass by a given urban block in the simulation are mapped to that urban block. This provides a way to look at the built environment and to identify specific locations that might link more strongly to given types of pedestrians or to overall numbers of pedestrians. Such analysis is useful in, among other things, identifying potentially public areas of the city and those that are more private.

Archaeological transportation analysis is a good example of one way new geospatial technologies can be used to better understand activities in the past. Without the geophysical and aerial survey data, the internal layout of the city would be unknown and unusable for simulations. Without the density and accuracy of data points provided by the micro-topographic GPS survey, or comparable data produced by different forms of LIDAR, too coarse a scale of resolution would yield largely meaningless results. Being able to accurately model changes in topography at an activity appropriate scale and link them to the physiology of human bodies allows for new avenues of hypothesis generation regarding social stratification or social dynamics within this city or within larger regions. Without these capabilities, none of this would be possible.

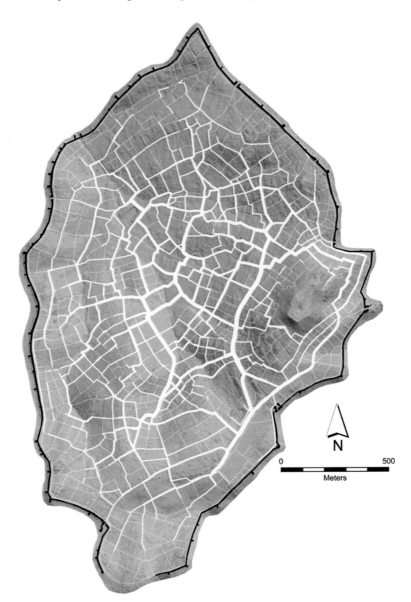

Fig. 19.3 Results of a simulation of female pedestrians moving from urban block to urban block along the streets of the city. The wider the width of the street line is, the greater the number of pedestrians choosing to use this street within the simulation

New advances in the use of technologies can create new problems to solve as well as new insights. Generating hypotheses is all well and good, but finding ways to test them on the ground can be challenging. In modern day situations, transportation analysis draws heavily upon observations of real-time movement to calibrate

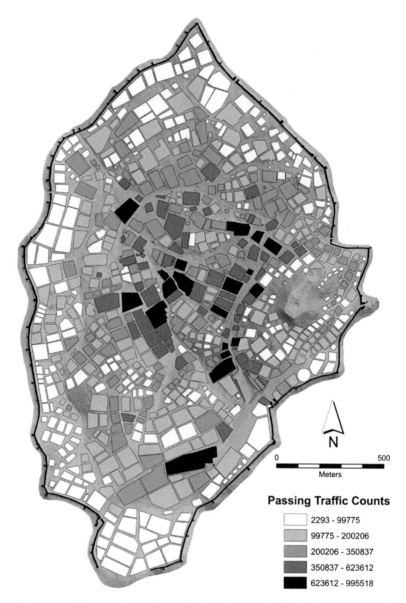

Fig. 19.4 Results of the same simulation of female pedestrians inside the city, with the total number of pedestrians passing by a given urban block or compound mapped to it. Differences in passing traffic are expected to reflect a private to public continuum for the different urban blocks as well as provide insight into areas that might have been in higher demand for particular portions of the population

and test models. For an ancient city that is not possible. Instead, additional technologies for analyzing micro-morphological and loose soil samples taken from excavated streets at Kerkenes Dağ were employed by the Charles McBurney

Laboratory for Geoarchaeology and the Department of Geography at Cambridge University in coordination with the project (Summers et al. 2005; Branting 2007, 2009). Analyses of compaction indicators in the micromorphology samples, as well as laser particle size analysis of the loose soil samples have proven useful as ways to identify relative levels of traffic that once used the streets. Initial correlation of this evidence with the output of the simulations suggests that these techniques can be used to independently measure ancient traffic.

Moving from the past, this work has a great deal of potential in the present as well. Simulating flows of pedestrians on a fine-scale is an area in Geography and Urban Planning that has been greatly underdeveloped (Litman 2007; Batty 2008). New simulation techniques are quite interesting for modeling pedestrian movement and could be paired with the methods already developed here for the management of cultural heritage sites (Zhou 2008; Torrens 2012). Being able to model flows of visitors between locations at a site, or between sites in a more regional perspective, would allow for the development of site infrastructure to maximize access while minimizing negative impacts. Work such as that of Shoval and Isaacson (2007), analyzing visitors movement in the old city at Akko in Israel using a time-geography framework, is of particular interest. It allows one to envision how this could be used for site management purposes. The transportation analysis was applied to a modern city using observations of pedestrians in the city center of Cambridge in England during its development phase (Branting 2004: 102–117). At Akko tracking data were created with a technological solution, GPS devices monitoring volunteer's movements, instead of direct observation (Shoval and Isaacson 2007: 288–290). An overarching modeling framework that combines both the simulations and the analysis of existing flows of movement would be a powerful tool for designing, analyzing and maintaining cultural heritage sites.

19.4 Conclusion

The future holds untold new technologies that will continue to alter the practice of archaeology and the management of our cultural heritage. Just as in the past, new technologies will create new opportunities to do things that were previously undreamt of. Some of these new advances will in time yield effective new methodologies for asking and answering critical questions for both the past and the present. Some advances will prove novel but of little practical utility. Yet one thing is as certain today as it was a century ago: technology is changing, and so should archaeology.

References

Adams, R. M. (1981). *Heartland of cities: Surveys of ancient settlement and land use on the central floodplain of the Euphrates*. Chicago: University of Chicago Press.

Batty, M. (2008). Fifty years of urban modelling: Macro-statics to micro-dynamics. In S. Albeverio, D. Andrey, P. Giordano, & A. Vancheri (Eds.), *The dynamics of complex urban systems: An interdisciplinary approach* (pp. 1–20). New York: Springer.

Bewley, R. H., & R czkowski, W. (2002). *Aerial archaeology: Developing future practice.* Amsterdam: Ios Press.

Bonabeau, E. (2002). Agent-based modeling: Methods and techniques for simulating human systems. *Proceedings of the National Academy of Sciences of the United States of America, 99,* 7280–7287.

Branting, S. (2004). *Iron Age Pedestrians at Kerkenes Dağ : An archaeological GIS-T approach to movement and transportation.* Ph. D. dissertation, Department of Anthropology, University at Buffalo.

Branting, S. (2007). Using an urban street network and a PGIS-T approach to analyze ancient movement. In J. T. Clark & E. M. Hagemeister (Eds.), *Digital Discovery: Exploring new frontiers in human heritage* (pp. 99–108). CAA 2006: Fargo, United States. Budapest: Archaeolingua.

Branting, S. (2009). Kerkenes Dağ project. In G. J. Stein (Ed.), *The Oriental Institute annual reports 2008–2009* (pp. 88–95). Chicago: Oriental Institute Press.

Branting, S. (2010). Agents in motion. In S. R. Steadman & J. C. Ross (Eds.), *Agency and identity in the ancient Near East: New paths forward* (pp. 47–59). London: Equinox.

Branting, S. (2012). Seven solutions for seven problems with least cost pathways. In D. A. White & S. L. Surface-Evans (Eds.), *Least cost analysis of social landscapes: Archaeological case studies* (pp. 209–224). Salt Lake City: University of Utah Press.

Branting, S., & Summers, G. D. (2002). Modelling terrain: The Global Positioning System (GPS) survey at Kerkenes Dağ, Turkey. *Antiquity, 76,* 639–640.

Branting, S., & Trampier, J. (2007). Geospatial data and theory in archaeology: A view from CAMEL. In R. B. Salisbury & D. Keeler (Eds.), *Space – Archaeology's final frontier? An intercontinental approach* (pp. 272–289). Cambridge: Cambridge Scholars Publishing.

Branting, S., Wu, Y., Srikrishnan, R., & Altaweel, M. R. (2007). SHULGI: A geospatial tool for modeling human movement and interaction. In M. North, D. Sallach, C. Macal (Eds.), *Proceedings of the agent 2007 conference on complex interaction and social emergence* (pp. 475–487). Argonne: Argonne National Laboratory.

Hägerstrand, T. (1970). What about people in regional science? *Papers of the Regional Science Association, 24,* 7–21.

Hillier, B. (2008). Space and spatiality: What the built environment needs from social theory. *Building Research & Information, 36*(3), 216–230.

Hillier, B., & Hanson, J. (1984). *The social logic of space.* Cambridge: Cambridge University Press.

Joyce, A. A. (2009). Theorizing urbanism in ancient Mesoamerica. *Ancient Mesoamerica, 20,* 189–196.

Kantner, J. (2012). Realism, reality, and routes: Evaluating cost-surface and cost-path algorithms. In D. A. White & S. L. Surface-Evans (Eds.), *Least cost analysis of social landscapes: Archaeological case studies* (pp. 225–238). Salt Lake City: University of Utah Press.

Kvamme, K. L. (1990). GIS algorithms and their effects on regional archaeological analysis. In K. M. Allen, S. W. Green, & E. B. W. Zubrow (Eds.), *Interpreting space: GIS and archaeology* (pp. 112–126). New York: Taylor & Francis.

Kvamme, K. L. (2003). Geophysical surveys as landscape archaeology. *American Antiquity, 68*(3), 435–457.

Litman, T. A. (2007). Economic value of walkability. *Transportation Research Record, 1828,* 3–11.

McDonald, I. (1961). Statistical studies of recorded energy expenditure of man: Expenditure on walking related to weight, sex, age, height, speed, and gradient. *Nutrition Abstracts and Reviews, 31*(3), 739–762.

Pandolf, K. B., Givoni, B., & Goldman, R. F. (1977). Predicting energy expenditure with loads while standing or walking very slowly. *Journal of Applied Physiology, 43,* 577–581.

Pred, A. (1986). *Place, practice and structure: Social and spatial transformation of southern Sweden*. Cambridge: Polity Press.

Pred, A. (1990). Context and bodies in flux: Some comments on space and time in the writings of Anthony Giddens. In J. Clark, C. Modgil, & S. Modgil (Eds.), *Anthony Giddens: Consensus and controversy* (pp. 117–129). London: Falmer Press.

Santee, W. R., Allison, W. F., Blanchard, L. A., & Small, M. G. (2001). A proposed model for load carriage on sloped terrain. *Aviation, Space, and Environmental Medicine, 72*(6), 562–566.

Shoval, N., & Isaacson, M. (2007). Sequence alignment as a method for human activity analysis in space and time. *Annals of the Association of American Geographers, 97*(2), 282–297.

Smith, M. L. (2008). Urban empty spaces: Contentious places for consensus-building. *Archaeological Dialogues, 15*(2), 216–231.

Summers, G. D., & Summers, F. (2010). From Picks to Pixels: Eighty years of development in the tools of archaeological exploration and interpretation, 1927–2007, at Kerkenes Dağ in Central Turkey. In P. Matthiae, F. Pinnock, L. Nigro, & N. Lorenzo (Eds.), *Proceedings of the 6th international congress on the archaeology of the ancient Near East* (pp. 669–683). Wiesbaden: Harrassowitz.

Summers, G. D., Summers, F., & Branting, S. (2004). *Megarons and associated structures at Kerkenes Dağ: An interim report* (pp. 7–41). XII: Anatolia Antiqua.

Summers, G., Summers, F., & Branting, S. (2005). *Kerkenes News, 7, 2004 – Kerkenes Haberler 7, 2004*. Ankara: METU Press.

Sun, J., Walters, M., Svensson, N., & Lloyd, D. (1996). The influence of surface slope on human gait characteristics: A study of urban pedestrians walking on an inclined surface. *Ergonomics, 39*(4), 677–692.

Torrens, P. M. (2012). Moving agent pedestrians through space and time. *Annals of the Association of American Geographers, 102*(1), 35–66.

Weyer, E. M., Jr. (1929). Exploring cliff dwellings with the Lindberghs. *The World's Work, 58*, 52–57.

White, D. A., & Barber, S. B. (2012). Geospatial modeling of pedestrian transportation networks: A case study from pre-Columbian Oaxaca, Mexico. *Journal of Archaeological Science, 39*, 2684–2696.

White, D. A., & Surface-Evans, S. L. (2012). *Least cost analysis of social landscapes*. Salt Lake City: University of Utah Press.

Zhou, Y. (2008). Agent-based modeling and simulation for pedestrian movement behaviors in space: A review of applications and GIS issues. *Proceedings of SPIE, 7143*(1), 714311-1–714311-9.

Chapter 20
Identifying Probable Archaeological Sites on Santa Catalina Island, California Using SAR and Ikonos Data

James C. Tilton and Douglas C. Comer

Abstract This chapter describes a method for identifying probable locations of archaeological sites over a wide area based on detecting subtle anomalies in vegetative cover through a statistically based analysis of remotely sensed data from multiple sources. This statistical analysis is further refined and elaborated to compensate for potential slight miss-registrations between the remote sensing data sources and the archaeological site location data. Data quantization approaches (required by the statistical analysis procedure) are explored, and a superior data quantization approach based on a unique image segmentation algorithm is identified. The effectiveness of the method is demonstrated with test data from Santa Catalina Island off the southern California coast.

Keywords Archaeology • Archaeological site detection • Archaeological survey • Archaeological modeling • Remote sensing in archaeology • Synthetic aperture radar • SAR.

20.1 Introduction

Comer and Blom (2007a, b, c) and Comer (2008) developed a means of utilizing aerial and satellite remote sensing data to identify probable locations of archaeological sites over a wide area based on detecting subtle anomalies in vegetative cover, soil characteristics, and rock scatters associated with sites on Santa Catalina Island, off the coast of Southern California. Remote sensing data included SAR (synthetic aperture radar) data collected by the AirSAR and GeoSAR (airborne synthetic aperture radar) platforms, and multiple band optical data from the Ikonos satellite sensor (operated by GeoEye, Inc.). A DEM (digital elevation model) was derived from the SAR data from which slope data was computed. These data, along with features computed from the Ikonos data served as input image layers for the anomaly detection approach.

D.C. Comer and M.J. Harrower, *Mapping Archaeological Landscapes from Space*, SpringerBriefs in Archaeology, DOI 10.1007/978-1-4614-6074-9_20, © Springer Science+Business Media New York 2013

In this chapter, we describe the remotely sensed data and the archaeological site data utilized for training and testing our analysis approach. We then explore data quantization approaches (required by our statistical analysis procedure), and we identify a superior data quantization approach based on a unique image segmentation algorithm. Next we provide a description of the statistical analysis and describe a further refinement and elaboration of this analysis to compensate for potential miss-registrations between the remote sensing data sources and the archaeological site location data. Finally, we demonstrate the effectiveness of the method with test data from Santa Catalina Island off the southern California coast.

20.2 Description of the Data and Necessary Preprocessing Steps

20.2.1 The Archaeological Site Data

The archaeological site location data was stored in ESRI shapefile (geospatial vector data) format with sites designated as "Lithic Scatter," "Habitation" and "Quarry." The tests in this study were performed using 36 habitation sites. The locations of these sites were collected in three field sessions conducted by author Douglas C. Comer with a crew of professional archaeologists and students: the first was from 11 August 2005 through 19 August 2005; the second from 26 December through 30 December 2005; and the third from 9 August through 14 August 2006.

20.2.2 The Remotely Sensed Data

The remotely sensed data employed in this study were collected by synthetic aperture radar (SAR) and multispectral sensors, the former carried by an airborne platform, the latter by a satellite platform.

20.2.2.1 Data from the Airborne GeoSAR System

For the Santa Catalina Island research, we utilized GeoSAR's X-band radar image data orthorectified to 3-m ground resolution (for more information on this system see http://www.geosar.com). This radar data was clipped at 1.2, and then scaled to integer values from 0 to 255. Clipping was necessary because of the well-known phenomena of radar speckle. Values of 1.2 and above generally indicate some level of speckle.

We also utilized DEM (digital elevation model) slope data computed from GeoSAR X-band radar data. The 3-m ground resolution orthorectification of the GeoSAR data was maintained in this derived slope data. The slope data ranges in integer values from 0 to 86. That is, slopes detected were from 0° to 86° (90° indicates a sheer cliff). The DEM slope data also included a mask indicating "no data" for non-land location (i.e., ocean), which were not used in our analysis.

20.2.2.2 Multispectral Data from the Ikonos Satellite and Features Derived from this Data

The Ikonos satellite provides 4-band multispectral (blue, green, red, and near infrared) data orthorectified to 4-m ground resolution (http://www.geoeye.com/CorpSite/). At 11-bit radiometric resolution (raw digital counts), this data ranges in integer values from 0 to 2,047. The Ikonos data utilized in this study data was a single scene collected on April 12, 2004 at 21:47 UTC. This Ikonos scene covers the central ~75% of the island.

We produced Tasseled Cap and Normalized Difference Vegetation Index (NDVI) features from the Ikonos data. We used the definition of Tasseled Cap that was specifically derived for Ikonos by Horne (2003), where it is noted, "Tasseled Cap transformations are linear transformations of the multispectral bands that allow easier identification of distinct surface types."

The NDVI feature is responsive to the presence and relative vigor of vegetation. We used the general definition of NDVI (see, for example Rouse et al. 1973).

While Horne's Tasseled Cap transform was specifically derived for Ikonos 11-bit raw digital count data, the NDVI transform is a general definition that is usually applied to calibrated radiance data, especially for cross sensor comparison or when relating NDVI to specific physical phenomena such as Leaf Area Index (e.g., see Soudani et al. 2006). However, since we are only using our NDVI feature as an indication of relative vegetation vigor, our calculation of NDVI from digital count data instead of calibrated radiance data is not an issue.

The Ikonos and Ikonos-derived features were all reprojected using nearest neighbor resampling to the 3-m pixel resolution of the DEM slope and X-band radar data.

20.2.2.3 Data Quantization Approaches

Our statistical analysis procedure requires that the data be quantized to no more than a few hundred integer values. The DEM slope and X-band radar data are already quantized in this manner. However, the 11-bit Ikonos data needs to be re-quantized to a smaller range of integer values. Since the Tasseled Cap and NDVI calculations yield floating point values, these feature values also need to be quantized to an appropriate range of integer values.

While histogram equalization or flattening (Rosenfeld and Kak 1976) may be appropriate for many image enhancement applications, it may not necessarily be the best quantization approach for statistical analysis, or even the best approach for enhanced visualization of image data. Inspired by earlier success in improved visual enhancement of medical images (Alexander et al. 2010), we also investigated a quantization approach based on the HSeg (Hierarchical Segmentation) image segmentation algorithm (Tilton et al. 2012). We utilized RHSeg, the recursive approximation of HSeg, to process the relatively large data sets.

HSeg is unique among image segmentation algorithms in being able to produce enhanced visualizations of images because of HSeg's unique tight integration of the aggregation of spatially disjoint region objects into region classes with the best merge region growing process. None of the other widely available image segmentation packages, such as ENVI EX, eCognition segmentation or ArcGIS Feature Analyst include this tight integration.

20.3   Analysis Approach and Results

The approach developed by Comer and Blom (2007a, b, c) and Comer (2008) is based on combining difference of means tests performed separately on single image layers to determine which sets of pixel values, found in image locations that coincide with archaeological sites, are statistically different enough to justify the assertion that they are obtained from a population different from the rest of the land area under study. The quantized pixel values for each single image layer are sampled within a meter buffer (m) drawn around the center point of known, randomly selected archaeological sites (n). These values are compared with quantized pixels values extracted from areas of the same size that were randomly selected from island areas that are at least $2*m$ meters distant from known archaeological sites. Our null hypothesis is that there is no difference between the population of values that lie within the prescribed buffer around the archaeological sites and the population of sites outside of these buffers. If the null hypothesis is upheld, it means that the quantized pixel values associated with the archaeological sites cannot contribute to distinguishing signatures for those sites. If, on the other hand, the null hypothesis is disproven, the quantized pixels values associated with these sites can be used to develop distinguishing signatures.

We tested the null hypothesis utilizing the standard Student's t-test procedure with a 95% threshold (T_{95}) based on degrees of freedom (dof). See, for example Tabachnick and Fidell (1996). We computed the dof for the case in which the two sets of samples may exhibit unequal variances (as is the case here). Given the dof value, the value of the T_{95} threshold for a two sided test can be looked up in a t distribution table, such as can be found at Crow et al. (1960).

We note that the Student's t-test cannot be performed with floating point values. Further, Student's t-test analysis will produce no positive detections if the range of quantized integer values is not limited to any more than a few hundred values. This is the motivation behind our study of alternate quantization approaches.

Including the derived Tasseled Cap and NDVI features, we considered the following 11 features:

Feature 1: DEM slope computed from the GeoSAR X-band radar data.
Feature 2: X-band radar data.
Feature 3: Band 1 (blue) of the Ikonos data.
Feature 4: Band 2 (green) of the Ikonos data.
Feature 5: Band 3 (red) of the Ikonos data.
Feature 6: Band 4 (near infrared) of the Ikonos data.
Feature 7: Tasseled Cap band 1 computed from the Ikonos data.
Feature 8: Tasseled Cap band 2 computed from the Ikonos data.
Feature 9: Tasseled Cap band 3 computed from the Ikonos data.
Feature 10: Tasseled Cap band 4 computed from the Ikonos data.
Feature 11: NDVI computed from the Ikonos data.

We re-quantized the Ikonos bands and Ikonos-derived features in four different manners in order to access how sensitive the analysis was to the quantization process. These features were re-quantized to 255 and 192 levels using histogram equalization and to 255 and 192 levels with the alternate re-quantization scheme utilizing HSeg or RHSeg.

The Student's t-test was performed separately on each of the 11 features using 15 known habitation sites (randomly chosen out of 36 known sites) and 15 randomly chosen locations on the Island. The data values considered to be associated with the site were the data values contained in a 12-m radius circle centered on each selected site location, which roughly corresponds to the average size of the archaeological sites under consideration. The degree of freedom was calculated for the case of unequal variances, and the T_{95} value was determined accordingly.

We then created an 11-band "positive hit" image with value 1 at locations where the quantized feature value corresponded to a digital number with a Student's t test score value $\geq T_{95}$ and zero otherwise. Using this "positive hit" image, we computed the "maximum positive hit density" for each of the known and unknown training sites used in the Student's t-test analysis. (We also computed the "maximum positive hit density" for the remaining known sites and an additional 15 unknown sites to serve as test sites.) This "maximum positive hit density" for a pixel is defined as the maximum of the positive hit densities computed over all circles of 12-m radius that include that pixel, where the positive hit density is defined as the number of pixels with value one in the positive hit image divided by the number of pixels in the 12 m radius circle centered on a pixel. This definition of maximum positive hit density is designed to compensate for 6–12 m of miss-registration.

The above was performed for ten cases of different random selections of known and unknown training (and test) sites. In each site selection case, the thresholds on the maximum positive hit density for each feature were exhaustively evaluated in which a known or unknown training or test site was considered to be detected if the maximum positive hit density for that site equaled or exceeded that threshold. A site was considered detected if it was detected for all features. The best set of thresholds were determined for all 11 features on the basis of maximizing the "training score" defined as the known training detection rate minus the unknown training detection rate. Then one feature was dropped in turn to determine which feature should be dropped in order to obtain the highest training score with ten features – and the remaining ten features were retained for further analysis. This continued until only one feature remained. Certain features appeared as "best features" more often than others. We found that features 1, 8, 10 and 11 occurred most frequently for the cases with the best training scores.

The analysis was then redone for all site selection cases starting from just features 1, 8, 10 and 11 (slope, Tasseled Cap band 2, Tasseled Cap band 4 and NDVI features). It was noted that the test scores were consistently poor for the ninth site selection case for all four quantization cases. So this case was dropped from further analysis.

The whole image was then analyzed for the best set of thresholds for all site selection case (except case 9). A habitation site was considered detected wherever the positive hit density equaled or exceeded the determined threshold for each particular feature (all features had to detect the site). Finally, the detections were summed over all nine runs.

20.4   Discussion

One indicator of the quality of performance for this type of analysis is the amount of consistency between the results produce from the different random selections of known and unknown training (and test) sites. We found that the 192 level RHSeg quantization results were clearly the most consistent across the nine random selections of known and unknown training (and test) sites. The 691 pixels detected across all nine sets of site selections with this quantization is nearly 24 times the 29 pixels detected across all nine sets in the 192 level histogram equalization quantization results, and over 138 times the 5 pixels detected across all nine sets in the 255 level histogram equalization quantization results. The 384 pixels detected across all nine sets in the 255 level RHSeg quantization results is also significantly more than the number of pixels (5 and 29) detected in either histogram equalization quantization result.

Table 20.1 displays the 192 level RHSeg quantization results along with proposed "concentration" and "gain" factors. These factors are a measure of how well the detection maps serve as maps of probable habitation sites that substantially concentrate the efforts of ground surveys to areas likely to contain habitation sites. Table 20.1 shows that the detection maps do provide significant concentration and gain factors. For example, for pixels detected as probable sites for six or more of the nine random selections of training (and test) sites, 36.11% of the known sites were detected with only 0.62% of the total land area being flagged as probable archaeological sites. This is a 58.48 times concentration of probable sites in the detected area, or a gain of 98.29%.

Table 20.1 Number and percentage (of total land area) of probable habitation site pixels detected and % known sites detected utilizing 192 level RHSeg based quantization plus proposed concentration and gain factors

Number of detections	Number of pixels detected	% detection (out of 12,873,517 pixels) (%)	% Known sites detected (%)	Concentration factor[a]	Gain[b] (%)
Nine or more	691	0.0054	2.78	517.51	99.81
Eight or more	7,304	0.057	8.33	146.88	99.32
Seven or more	31,663	0.25	19.44	79.06	98.74
Six or more	79,499	0.62	36.11	58.48	98.29
Five or more	153,967	1.20	52.78	44.13	97.73
Four or more	260,199	2.02	66.67	32.98	96.97
Three or more	410,799	3.19	88.89	27.86	96.41
Two or more	639,730	4.97	97.22	19.56	94.89
One or more	1,162,841	9.03	100.00	11.07	90.97

[a]Concentration factor = % known sites detected/% detection
[b]Gain = 1 − (% detection/% known sites detected)

Fig. 20.1 (a) An RGB representation of some of the analyzed image features: *Red* = TC$_2$, *Green* = TC$_4$, and *Blue* = NDVI. All are displayed with 255 level RHSeg quantization. (b) Colored coded display of the number of positive detections of probable habitation sites: 1 detection = *blue*, 2 detections = *magenta*, 3 detections = *cyan*, 4 detections = *green*, 5 detections = *yellow*, and 6 or more detections = *red*

Figure 20.1a provides a RGB representation of three of the image features analyzed. We selected three of the features that proved most useful in our analysis: Tasseled Cap band 2 (displayed as red), Tasseled Cap band 4 (displayed as green) and NDVI (displayed as blue). All features are displayed with 255 level RHSeg quantization.

Figure 20.1b displays a map of detections of probable habitation site locations for the 192 level RHSeg quantization results (as in Table 20.1). The red locations correspond to six or more detections out of the nine test runs.

We have described an approach for identifying probable archaeological sites from remotely sensed data. The described approach improves on earlier work through the inclusion of a method for compensating for potential slight miss-registrations between the remote sensing data sources and the archaeological site location data used for training the approach. A further improvement is utilizing a unique data quantization approach based on image segmentation for providing the data quantization required for our Student's *t*-test based analysis approach. This unique data quantization approach is shown to give superior results when compared to conventional histogram equalization quantization. We hope that the probable site maps produced through this approach, such as Fig. 20.1b, will be of great assistance to ground survey teams in finding vulnerable archaeological sites before they are compromised or destroyed.

Acknowledgments This work was performed under the project "Automating and Enhancing Protocols for the Development of Signatures for Archaeological Sites Using Publicly Available NASA Imagery," supported by grant number 07-SAP07-0013 from NASA's Space Archaeology program.

References

Alexander, S., Gran, R., & DeWitt, S. (2010). NASA technology may aid interpretation of medical imagery, NASA Press Release 10–261: http://www.nasa.gov/home/hqnews/2010/oct/HQ_10-261_NASA_Imagery.html. Accessed 7 Nov 2012.

Comer, D. C. (2008). Wide-area, planning level archaeological surveys using SAR and multispectral images. *Proceedings of the geoscience and remote sensing symposium, 2008. IGARSS 2008* (pp. 45–47). New York: IEEE International.

Comer, D. C., & Blom, R. G. (2007a). Detection and identification of archaeological sites and features using synthetic aperture radar (SAR) data collected from airborne platforms. In J. R. Wiseman & F. El-Baz (Eds.), *Remote sensing in archaeology* (pp. 103–136). New York: Springer Science + Business Media.

Comer, D. C., & Blom, R. G. (2007b). Remote sensing and archaeology: Tracking the course of human history from space. Earth Imaging Journal, March/April 2007, pp. 10–13.

Comer, D. C., & Blom, R. G. (2007c). Wide area inventory of archaeological sites using aerial and satellite data sets: Prologue to resource monitoring and preservation (with Ronald G. Blom). *Proceedings of the 32nd international symposium on remote sensing of environment,* San Jose, 25–29 June, (pp. 452–456). Ann Arbor: Environmental Research Institute of Michigan.

Crow, E. L., Davis, F. A., & Maxfield, M. W. (1960). *Statistics manual.* New York: Dover Publications.

Horne, J. H. (2003). A tasseled cap transformation for IKONOS images. *Proceedings of ASPRS 2003 conference, Anchorage*, 5–9 May. Bethesda: American Society of Photogrammetry and Remote Sensing.

Rosenfeld, A., & Kak, A. C. (1976). *Digital picture processing*. New York: Academic.

Rouse, J. W., Haas, R. H., Schell, J. A., & Deering, D. W. (1973). Monitoring vegetation systems in the great plains with ERTS. *Proceedings of the third ERTS symposium*, 10–14 Dec, (pp. 309–317). Greenbelt, MD: Goddard Space Flight Center.

Soudani, K., François, C., le Maire, G., Le Dantec, V., & Dufrêne, E. (2006). Comparative analysis of IKONOS, SPOT, and ETM + data for leaf area index estimation in temperate coniferous and deciduous forest stands. *Remote Sensing of Environment*, 102, pp. 161–175.

Tabachnick, B. G., & Fidell, L. S. (1996). *Using multivariate statistics*. New York: HarperCollins.

Tilton, J. C., Tarabalka, Y., Montesano, P. M., & Gofman, E. (2012). Best merge region growing segmentation with integrated non-adjacent region object aggregation. *IEEE Transactions on Geoscience and Remote Sensing*, 50(11), pp. 4454–4467.

Chapter 21
Refinement of a Method for Identifying Probable Archaeological Sites from Remotely Sensed Data

Li Chen, Douglas C. Comer, Carey E. Priebe, Daniel Sussman, and James C. Tilton

Abstract To discover and locate archaeological sites, we aim to develop scientific and efficient approaches to identify these sites with high accuracy. In this chapter, we present a statistical learning model consisting of imaging processing, feature extraction and classification. Our analysis uses the remotely sensed data composed of eight WorldView-2 imagery bands and one slope band. In the imaging processing step, we use a particular annuli technique; in the feature extraction step, principal component analysis is applied; in the classification step, linear discriminant analysis is carried out. We test this procedure on 33 lithic sites, 16 habitation sites and 100 non-sites from the western portion of Ft. Irwin, CA, USA. The receiver operating characteristic curve, used for assessing the performance of the algorithm, shows that our new approach generates higher classification power than the Archaeological Predictive Model (APM). When APM is convexly combined with our new model, the classification accuracy is even higher.

Keywords Imaging processing • Feature extraction • Classification

Archaeological sites are being destroyed or compromised at a catastrophic rate in most regions of the world in this age of globalization. To facilitate locating archaeological sites and regions free of archaeological sites, we developed a statistical classification approach that leverages differences in vegetative cover, soil chemistry, and soil moisture by analyzing remotely sensed data from multiple sources at locations with and without archaeological significance. For some uses, it is important to find regions without archaeological sites so that these areas can be used for other purposes. We report here on an analysis of the western portion of Ft. Irwin, CA, USA that uses 8-band multispectral Worldview-2 imagery and slope data computed

D.C. Comer and M.J. Harrower, *Mapping Archaeological Landscapes from Space*,
SpringerBriefs in Archaeology, DOI 10.1007/978-1-4614-6074-9_21,
© Springer Science+Business Media New York 2013

from a synthetic aperture radar (SAR) derived digital elevation model (DEM). We first describe the remotely sensed data and the archaeological site data that was used for training and testing our approach. We then describe our new technique, and demonstrate the effectiveness of our approach on the Ft. Irwin data set and compare our results with an Archaeological Predictive Model (APM) (Ruiz 2003).

The remotely sensed data employed in this study was multispectral WorldView-2 data. DigitalGlobe, Inc., Longmont, CO, USA (http://www.digital-globe.com) contributed 14 swaths of WorldView-2 satellite imagery data to our study covering the entire land area of the China Lake and Ft. Irwin military reservations in California, USA. We report on our use of just one of these satellite imagery data swaths, collected on December 30, 2010, covering a large part of the western portion of Ft. Irwin. While DigitalGlobe provided both panchromatic and multispectral data, we utilized only the multispectral data in our study. The data is orthorectified to 2-m ground resolution and has 11-bit radiometric resolution (stored as 16-bit integers). DigitalGlobe provided this swath of data in 45 separate 4096×4096 pixel blocks of data, which we mosaicked together to form an eight spectral band data set with 9,859 columns and 61,098 rows. We used in our analysis a 9,859 column by 23,000 row subset of this data that covered the grounds of a western portion of Ft. Irwin. All locations outside of the Ft. Irwin land area boundary were masked out, seen in Fig. 21.1. The slope data used in this study was derived from SAR data from the airborn GeoSAR system. As described on the GeoSAR website (http://www.geosar.com/):

> As the world's only dual-band, single-pass interferometric synthetic aperture radar (IFSAR) mapping system, GeoSAR concurrently collects both surface features and bare-earth elevation data using X- band and P-band radar.

The slope data was derived from a DEM computed from GeoSARs X-band radar image data orthorectified to 2-m ground resolution. The slope data ranges in integer values from 0 to 86. That is, slopes detected were from $0°$ to $86°$ ($90°$ indicates a sheer cliff). We utilized all possible combinations of band difference ratios (Marchisio et al. 2010) of the eight multispectral bands and slope as features for our study. Band difference ratios offer some robustness to changes in lighting conditions at different times of data collection, which will help in our eventual extension of our analysis to other portions of Ft. Irwin and to the nearby China Lake site. These band difference ratio features are defined as

$$B_{ij} = \frac{B_i - B_j}{B_i + B_j}, \text{ for all } i > j. \tag{21.1}$$

Archaeological predictive models (APMs) have been used in archaeology for several decades now to predict areas in which archaeological sites are likely to be found. Rather than finding the sites themselves, APMs are intended to identify areas

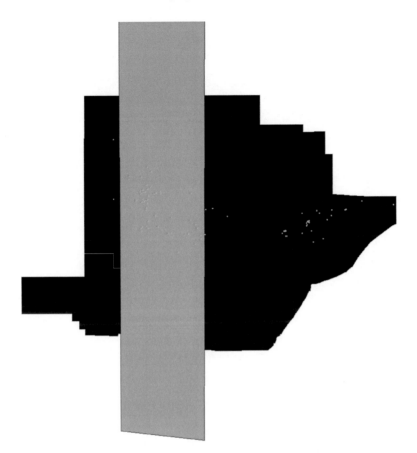

Fig. 21.1 WorldView-2 data Black:Ft. Irwin. Green:swath we analyzed in paper

within which sites might be found and deal only with how suitable an area is for a certain activity. The APM for Ft. Irwin (Ruiz 2003) was composed of several individual APMs, each predicting areas in which different types of archaeological sites might be found. We will consider only the APMs done for two types of sites: habitation and lithic. An APM for these habitation sites is seen in Fig. 21.2. Areas in green are those which are predicted to be most favorable for the presence of archaeological sites, red the least, and colors between those two in order of decreasing likelihood. ROC curves of APM presented in Fig. 21.3 compare the performances of the APM and the site detection protocol that we have developed. It is significant that the two approaches can be combined to produce a better performance than could be obtained by either alone.

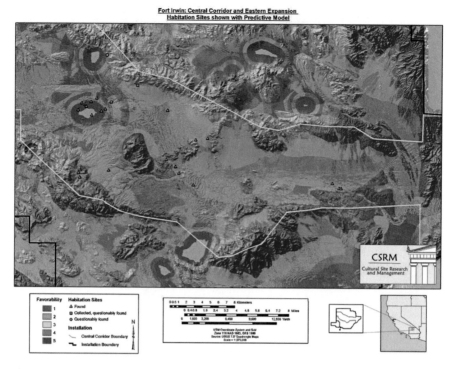

Fig. 21.2 The archaeological predictive model for habitation sites. (**a**) Habitation (**b**) Lithic

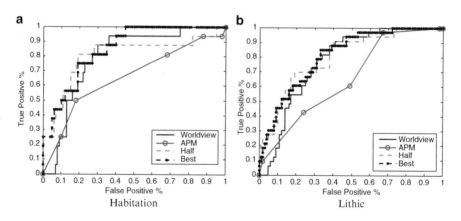

Fig. 21.3 ROC curves for various choices of γ

Our statistical classification scheme consists of three steps: image processing f, feature extraction g and classification h. We consider our data as Z two-dimensional arrays that we refer to as feature images. Consider that each $s = (s_1, s_2) \in S = \{1, 2, \dots, w\} \times \{1, 2, \dots, h\}$ corresponds to a specific geographical location. A feature image $B \in \mathbb{R}^{w \times h}$ maps each location $s \in S$ to a real-valued intensity. Additionally, each site location is labeled either 1, indicating the site has archaeological significance, or 0 otherwise. We seek to determine these labels, which are known for only part of the region, from Z feature images.

Given the setup and notations, the image processor, $f : S \times \left(\mathbb{R}^{w \times h}\right)^{z} \rightarrow \mathbb{R}^{\tilde{d}}$, takes a location and Z feature images, and transforms them into a vector of statistics. For each location s,

$$f\left(s, B^{(1)}, \ldots, B^{(Z)}\right) = \begin{pmatrix} f_1\left(s, B^{(1)}\right) \\ f_2\left(s, B^{(2)}\right) \\ \vdots \\ f_B\left(s, B^{(Z)}\right) \end{pmatrix} \in \mathbb{R}^{\tilde{d}} \ \forall s, B^{(1)}, \ldots, B^{(Z)} \qquad (21.2)$$

We assume $f_z : S \times \mathbb{R}^{w \times h} \rightarrow \mathbb{R}^{\tilde{d}/Z}$ depends only on image z for each image and all f_z are identical. After image processing, each site is represented by a vector in $\mathbb{R}^{\tilde{d}}$. Due to high dimensionality of the feature space and comparably low sample size, a feature extractor $g : \mathbb{R}^{\tilde{d}} \rightarrow \mathbb{R}^{d}$, with $d \ll \tilde{d}$, is used for lowering the feature dimensions while capturing the most variation, eliminating irrelevant features, and denoising the dataset (Duda et al. 2000; Tan et al. 2006). After image processing and feature extraction, the classifier $h : \mathbb{R}^{d} \mapsto \{0,1\}$ assigns class labels $\{0, 1\}$ to each feature extracted data vector. Thus, our classification algorithm $h \circ g \circ f$ is complete. Both the feature extraction dimension and the classifier parameters are chosen data-adaptively.

We use a particular imaging processing method: annuli technique, to obtain statistics from sites. The annuli technique provides rotation-invariant analysis. If it were not valid, the bias-variance tradeoff in data modeling would cause failure. For this step, define $A_s\left(r^{(in)}, r^{(out)}\right) = \left\{s' \in S : r^{(in)} \leq \|s - s'\| < r^{(out)}\right\}$ to be the set of coordinates in the annulus centered at s with inner radius $r^{(in)}$ and outer radius $r^{(out)}$. We calculate the robust statistics median and median absolute deviation (MAD) for the pixel intensities in each annulus. Our method considers three annuli sizes with radii differences at $\{2, 4, 6\}$, where there are ten annuli for each size and thus 30 annuli in total. Thus, the annuli method transforms each location to a vector in $R^{2 \times 30 \times Z}$, where Z denotes the number of feature images, which in this case is the 36 band difference ratios.

The feature space hence has $\tilde{d} = 2,160$ dimensions, far larger than the sample size 116 for habitation and 133 for lithic sites. The next step, feature extraction, is specifically fulfilled by principal component analysis (PCA) to lower the dimensions to $d \ll \tilde{d}$. To select the PCA dimension d we perform a leave-one-out estimate of error for each selection of d from 1 to n-1, and choose the new dimension d associated with the smallest error. The new feature space is thus $\mathbb{R}^{d}$ and the linear discriminant analysis (LDA) classifier is applied in $\mathbb{R}^{d}$.

The classifier LDA is the Bayes plug-in classifier under the assumption that the data is distributed according to class conditional multivariate normal distributions with the same covariance matrices. After estimating the means and covariances, we compute estimated posterior probabilities, $\hat{\eta}(x)$, the probability that a site with extracted feature vector x has label 1. The LDA classifier h is defined as $h(x) = 1\{\hat{\eta}(x) > \tau\}$, i.e., the site has archaeological significance if the estimated posterior probability is greater than the threshold $\tau \in [0,1]$.

The above analysis was used on 36 feature images which correspond to the band difference ratios computed from all pairs of the 8-spectral band WorldView-2 imagery combined with the slope data. We have resorted to the unconventional step of including slope via band difference ratios because we have found that doing so produces significantly better results than including the slope data as a single additional feature; this phenomenon deserves further investigation. For our true sites we consider two different classes, lithic sites ($n_1 = 33$) and habitation sites ($n_1 = 16$). We trained separate classifiers for the lithic and habitation classes. For the class zero training data we selected $n_0 = 100$ locations uniformly at random from the surveyed region of the image swath. We also ensured that the selected locations were at least 100 pixels from every known true site.

The result of our analysis is a posterior probability for each site derived from the classifier trained on the remaining sites. We denote this posterior by $\hat{\eta}_{wv}$ for the WorldView derived posterior. Independent of the WorldView analysis, the archaeological predictive model (APM) provides a score which, when normalized to {0, 0.2, 0.4, 0.6, 0.8, 1}, can be considered as an alternative posterior estimate which we denote $\hat{\eta}_{APM}$. We consider convex combinations of the APM posterior and the Worldview analysis posterior. For each value of $\gamma \in [0,1]$ this gives a new posterior estimate:

$$\hat{\eta}_\gamma = (1 - \gamma)\hat{\eta}_{APM} + \gamma\hat{\eta}_{wv} \qquad (21.3)$$

Figure 21.3 shows the receiver operating characteristic (ROC) (Duda et al. 2000) curve of lithic sites and habitation sites corresponding to four classifiers: APM $(\gamma = 0)$, Worldview-2 $(\gamma = 1)$, half of each $(\gamma = 0.5)$, and the optimal combined model. For different choices of false positive percentage (or true positive percentage) we see that the combined model typically improves performance over using either the APM and the WorldView analysis alone. The consistent superior performance of the combined model to APM, indicates that including our method increases prediction power over using the APM alone. Figure 21.4 presents comparative results for APM vs. WorldView for lithic sites at false positive rate 0.49. We see that the false negative rate is much lower in our method than in APM. The confusion matrices for each classification scheme are below.

	WorldView		APM	
	True	False	True	False
True	31	2	20	13
False	49	51	49	51

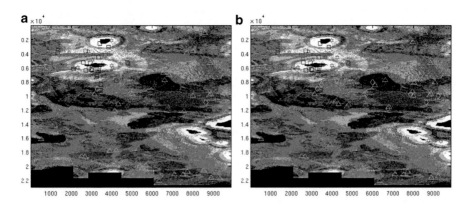

Fig. 21.4 Comparing results for our analysis as compared to the APM. The background is colored according the Lithic APM values. Legend: □ True Positive, ◊ False Negative, △ True Negative, * False Positive. (**a**) Map of Lithic sites with WorldView classification. (**b**) Map of Lithic sites with APM classification

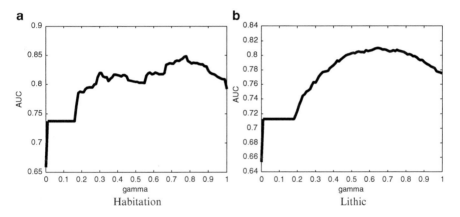

Fig. 21.5 Area under the ROC curve for combined model with respect to γ. (**a**) Habitation (**b**) Lithic

Figure 21.5 shows the area under the ROC curve (AUC) (Duda et al. 2000), a performance summary statistic, for each $\gamma \in [0,1]$ where 0 represents using only the APM and 1 represents using on the WorldView-2 analysis. For the habitation sites, Fig. 21.5a, we see that the WorldView dominates APM and that a combined classifier with $\gamma \approx .78$ provides even better performance. Noticeably, there is a jump in the AUC at $\gamma = 0$. This is due to the fact that the APM gives only six possible values so for γ near zero the WorldView analysis acts as a tie breaker for sites that have the same APM score. Similar behavior is see for lithic sites, Fig. 21.5b, the WorldView analysis outperforms the APM and that the a combined classifier with $\gamma \approx .65$ provides even better performance.

Our classification framework, specified by the three components: image processor, feature extractor and classifier, is one example of a statistical approach to archaeological site discovery. We compared our method to an archaeological predictive model (APM), which has been considered as the standard means for identifying areas that are likely to contain archaeological sites, and demonstrated the effectiveness of our new approach over a portion of the Ft. Irwin military reservation in California. Almost certainly, we can further improve performance by considering alternative methods in each of the three steps in our classification framework. For the imaging processing step, if we incorporate more known archaeological properties for the sites, we can use different image processors for remote sensing. We used PCA composed with LDA, where there is a wealth of empirical evidence suggesting that composition of the two procedures gives improved results over merely using LDA on the original feature space. One may also use a different feature extractor and alternative classifiers such as support vector machine, boosting, and k-nearest neighbor. In addition to changing any of the three components, including additional feature bands or data possibly related to site properties will improve classification power.

We have reported here a statistical treatment for predicting archaeological site locations and showed that combining our approach with APM yields improved performance. We plan to extend our work to other portions of Ft. Irwin and nearby China Lake military reservation, with the ultimate objective of establishing the capacity to conduct rapid, wide-areas surveys for archaeological sites with reliable results. We are optimistic that we can develop more productive and efficient methods for finding archaeological sites in the future.

References

Duda, R.O., Hart, P. E., & Stork, D. G. (2000). *Pattern classification*, 2nd ed. Wiley-Interscience. New York.

Marchisio, G., Padwick, C., & Pacifici, F. (2010). Evidence of improved vegetation discrimination and urban mapping using worldview-2 multi-spectral imagery. *ASPRS annual conference*. Baltimore, Maryland.

Ruiz, M. O. (2003). The development and testing of an archaeological predictive model. Technical report, U.S. Army Corps of Engineers Construction Engineering Research Laboratory, Champaign.

Tan, P.-N., Steinbach, M., & Kumar, V. (2006). *Introduction to data mining*. Boston: Pearson Addison Wesley.

Chapter 22
Survey, Automated Detection, and Spatial Distribution Analysis of Cairn Tombs in Ancient Southern Arabia

Michael J. Harrower, Jared Schuetter, Joy McCorriston, Prem K. Goel, and Matthew J. Senn

Abstract Small circular cairn tombs found across southern Arabia offer an intriguing window into the movements and mortuary practices of ancient hinterland peoples. This chapter briefly reports preliminary results of archaeological survey, automated detection efforts and spatial distribution analysis for such tombs across the Southern Jol area of Hadramawt Governate Yemen. The most detailed level of survey recording was conducted for 394 tombs, an automated detection algorithm (to shortly be reported in far greater detail elsewhere) yielded highly promising results, and GIS analyses indicate associations between tombs and water-rich areas.

Keywords Yemen • Oman • Arabia • Mortuary archaeology • Tombs • Archaeological survey • Satellite remote sensing • GIS • GPS

22.1 Introduction

Cairn tombs are among the most visibly prominent remains left by hinterland peoples of ancient Southern Arabia. In parts of Yemen and Oman even the causal traveller is hard pressed not to notice such small circular tombs variably found and referred to over the years by explorers and archaeologists as turret graves, pillbox cairns, or Hafit tombs (e.g. de Cardi et al. 1976; de Maigret 1996; Frifelt 1975) that bear an approximate resemblance and perhaps shared ancestry with comparable third millennium B.C. examples to the north including as far away as the Sinai and Syria (Steimer-Herbet 2004; Yule and Weisgerber 1998).

These cairn tombs (Fig. 22.1) that we hereafter refer to as High Circular Tombs, or HCTs, are arguably the most abundant hinterland landmarks across the mountainous highlands of Yemen and Oman, yet they certainly are not evenly distributed across landscapes and are not found everywhere (Deadman 2012; Giraud 2010). Closer attention also reveals numerous other types of small monuments including linear arrangements of three clustered upright stones known as triliths, wall tombs, dolmens, anthropomorphic statue/menhir stelae, and madhba grilling hearths (Bin'Aqil and McCorriston 2009; McCorriston et al. 2011). These tombs and monuments are far more than merely places to build a fire or dispose of the dead: they are lasting manifestations of ancient tribal affiliations expressed through spatial and temporal variability in style and form. They not only mark tribes-peoples movements, but through the idiom of kinship mark presence of ancestors, convey social identities and assert social affiliations.

The many thousands of HCT tombs across the region are often strung along the cliff lines above wadi drainages (Fig. 22.2) and their distribution includes more than a thousand clustered at each of the massive necropolises of Jebel Jidran and Jebel Ruwaik in Yemen's Ramlat as-Sabat'ayn Desert interior (Steimer-Herbet 1999, 2001). Their abundance, wide distribution, and relative inaccessibility make physically visiting and comprehensively mapping them all over large areas highly challenging and ineffective. Remote sensing (RS) and GIS and GPS technologies are thus exceptionally well-suited to survey some of the tombs in different areas, detect others in imagery, and subsequently model their spatio-temporal distribution providing a window into the lives of ancient hinterland peoples and their role in the genesis of early complex polities.

The Roots of Agriculture in Southern Arabia (RASA) project initiated by Joy McCorriston in 1998 began with a concentration on the beginnings of agriculture and initially recorded tombs only occasionally when they happened to coincide with the search of stratified deposits that might hold evidence of early agriculture suitable for excavation. After obtaining high (0.6 m) spatial resolution Quickbird satellite imagery in summer 2004 to map an area of particular interest, we quickly realized we could identify and map tombs from imagery avoiding the arduous task of climbing the multitude of high cliffs to directly map them by hand. For pilot study fieldwork concentrating on HCTs in winter 2005, we manually identified and marked tombs visible on Quickbird imagery. It rapidly became clear this procedure was not only time consuming but enormously subjective. Even for a single human observer, from hour-to-hour and day-to-day it was very difficult to adhere to definitive criteria while systematically searching images over different types of terrain (cf. Deadman 2012).

We therefore assembled a collaborative team, The Ancient Human Social Dynamics (AHSD) Project, concentrating on tombs and other small-scale monuments involving faculty and students in geodetic science, spatial statistics, and anthropology. Our approach involved archaeological survey and excavation, automated detection of monuments in high resolution satellite imagery, ground-truthing, and resultant spatial analyses. The AHSD team's fieldwork concentrated in Hadramawt Governate of eastern Yemen in winter 2008 (Fig. 22.3) and Dhofar Province of western Oman in winter 2009 and 2010. This chapter offers a brief overview of field survey data, detection, and preliminary GIS analysis efforts for the Yemen dataset (see McCorriston et al. 2011); a detailed reporting of automated detection methods (Schuetter et al. in review) and results for Oman (Harrower et al. in press; McCorriston et al. in press) will be shortly forthcoming elsewhere.

Fig. 22.1 High Circular Tombs (HCT) in Hadramawt Governate (Wadi Idm *right*, Wadi Sana *left*)

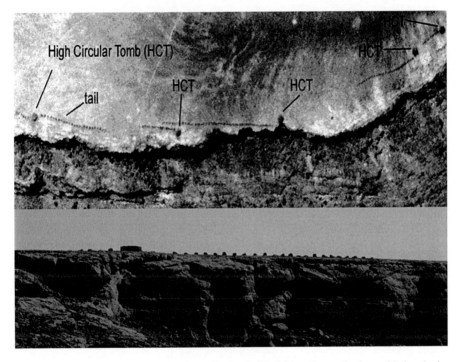

Fig. 22.2 High Circular Tombs along a cliff edge in Wadi Idm (the tomb pictured below is the leftmost tomb in the panchromatic Quickbird image above)

22.2   Archaeological Survey and Data Collection for Detection and GIS Analysis

Building on previous field experience since 1998 in the rugged highlands of Hadramawt's Southern Jol (McCorriston et al. 2002, 2005) including RS/GIS/GPS focused efforts (Harrower 2008a, b; Harrower et al. 2002, 2012) the AHSD team

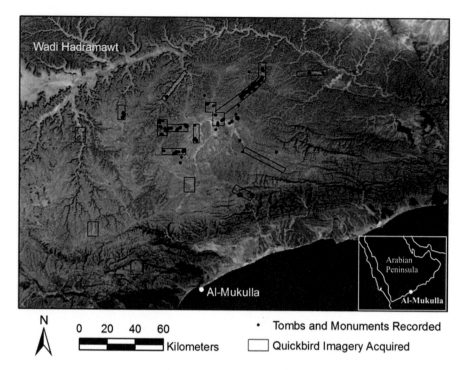

Fig. 22.3 The AHSD Project study area including tombs and monuments recorded during 2008 survey and areas of acquired Quickbird imagery

devised a plan tailored to the challenging terrain with very few maintained roads. Following on our initial 2004 acquisition of Quickbird (QB) we purchased ten additional QB images prior to fieldwork in areas where we anticipated substantial numbers of tombs based on previous experience, and six more images following fieldwork in areas where we collected considerable field data on tombs and other monuments (for a total of 1,350 km² of QB imagery). We utilized a kinematic GPS configuration consisting of four Trimble 5700 receivers (one base station, occasionally a sub base station, and two rovers) to ensure better than half meter accuracy so our GPS data could be accurately overlaid on high resolution imagery. We specifically targeted a wide range of areas on different landforms, in different contexts to generate a sample of monuments across a very large approximately 20,000 km² area, which could therefore help illuminate regional spatial patterning of tombs and other monuments.

In sum, though our movements were substantially curtailed because unanticipated security concerns, we recorded more than 30,000 GPS positions, yet only a fraction of these are our focus. The most detailed level of recording – consisting of a GPS center point, tomb/monument boundaries, a paper tomb/monument form, and photographs – was completed for 394 HCT tombs, 47 wall tombs, and 37 small monuments and structures designated in our "other" category. A reduced level of recording consisting of a GPS center point and photographs was conducted for less well-preserved tombs and monuments for which there was uncertainty as to their nature, including cases deemed possible HCTs and a multitude of adjacent features such as stone piles, stone rings, and hearths.

22.3 Automated Detection: A Preliminary Description

The process of devising procedures for automated detection of HCT tombs was challenging, complex and was the central subject of a pioneering Ph.D. dissertation in statistics completed by AHSD team member Schuetter (2010). Since algorithm details will be presented in far greater detail elsewhere (Schuetter et al. in review) we provide only a brief description below. Even though many tombs can been seen on QB imagery with the naked eye, programing a computer to identify them involves a wide range of obstacles both in terms of supplying training and subject imagery and implementing iterative procedures that can accurately distinguish tombs from objects of similar size and appearance (including trees, boulders, circular rock outcrops and discolorations). Generating training sample data involves providing precise image coordinates for identifiable, well-preserved tombs. Quickbird imagery (the highest resolution imagery available at the project's outset) consists of five bands from 60 cm (band 1) to 2.4 m (bands 2–5) spatial resolution. While the precision of our GPS data is less than 10 cm, lacking a high-spatial resolution DEM, our imagery rectification efforts have yet to achieve better than a few meters RMSE (Roots Mean Squared Error) accuracy (Digital Globe Inc. 2012). While we hope to eventually resolve this issue as more advanced rectification methods become available, training set coordinates have thus far been manually obtained by comparing GPS records with imagery and making required adjustments accordingly.

Automated detection necessarily involves iterative procedures that sequentially eliminate candidate objects. The initial detection approach, implemented in Matlab software, involves five stages that evaluate: (1) if an object is present, (2) if the object is vegetation, (3) if the object is the correct size, (4) if the object is circular, and (5) if the object's multivariate attributes are similar to those of other HCTs (Schuetter et al. in review). Our surveyed dataset of HCTs in Yemen includes tomb diameters from 1.3 to 9.0 m (with an average of 3.72 m). We therefore initially determined that a 25×25 pixel outer moving window generating data for a potential HCT and its immediate surroundings, combined with an inner 5×5 pixel moving window generating data for a potential HCT itself offered a promising first step to determine presence/absence of a candidate object. The second step relies on the longstanding Normalized Difference Vegetation Index (NDVI) based on vegetation's differential reflectance/absorption of infrared versus red light. This stage helps eliminate trees and other vegetation that can often appear similar to HCTs (Fig. 22.4). Although only the first

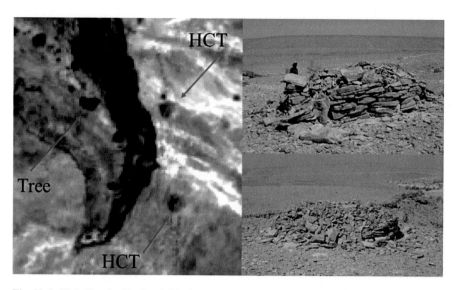

Fig. 22.4 High Circular Tombs visible in pan-sharpened 60 cm Quickbird imagery (the tombs in the *left* image are pictured *right* and can be distinguished in part because of pronounced contrast with vegetation)

QB band is 60 cm resolution, pan-sharpening was used to merge or sharpen the first panchromatic QB band with the other four QB bands, creating a five band image of 60 cm spatial resolution (Fig. 22.4).

The third step further evaluates the size of the candidate object. In general, the boundary of an HCT is often marked by an abrupt shift in pixel values from inner darker values to outer lighter values. Using consecutive 3×3, 5×5, and 7×7 pixel nested windows, this third step evaluated when the inner window surpassed the median value for the entire 25×25 pixel outer window. Since HCTs tend to be more circular than other natural objects, the fourth stage of detection evaluates the circularity of candidate objects using, in part, a Hough transform circle fitting and boundary extraction. Objects lacking sufficient circularity are thus removed as candidates. The final, fifth stage of the initial detection algorithm considers candidate objects' multivariate intensity, contrast, shape, and NDVI attributes in a six-dimensional feature space. Candidates outside the convex hull of six-dimensional feature space for the training set can thus be removed from the list of potential HCTs.

While the aforementioned detection procedures continue to be developed and refined preliminary quantitative evaluations yielded 50–88% success rate in detecting test HCTs in six test runs with between 13,000 and 38,000 false positives per run (Schuetter et al. in review). Although still at an incipient stage, we anticipate this research will help lay the foundations for increasingly efficient and accurate means of tomb and archaeological feature detection.

22.4   GIS Analysis of Association Between HCT Tombs and Water Flow

As monuments that pronounce the landscape presence of mobile nomads, HCT tombs are visually powerfully symbols that one would expect might be distributed in relation to important resources and territories. Interestingly, HCT tombs first appear during the late fourth millennium, perhaps associated in space and time with the region's earliest documented agriculture and irrigation (Deadman 2012; Giraud 2010; Harrower 2008a, b).

We applied GIS analysis to evaluate possible associations between HCT tombs and water-rich areas. Our survey was in many respects exploratory and aimed to gather data on as many tombs and monuments as quickly as possible to facilitate development of automated detection. Since no fixed boundary limited our field data collection, we used the directional distribution function of ArcGIS software (rather than a subjectively defined arbitrarily delineated boundary) to establish a sample universe area to analyze the distribution of HCT tombs. The directional distribution function "creates standard deviational ellipses to summarize the spatial characteristics of geographic features: central tendency, dispersion and directional trends." (ESRI (Environmental Systems Research Institute 2011). Using the third standard deviation setting, this function created an ellipse covering 24,060 km^2 that encompasses our surveyed dataset (Fig. 22.5). A Shuttle Radar Topography Mission (SRTM) flow accumulation layer quantified the relative abundance of water flow based on topography (Harrower 2010; Maidment 2002). Using the focal statistics function, this flow accumulation layer was averaged to create a new layer depicting mean flow accumulation within a 500 m radius. Five hundred meter radius mean flow accumulation values were then extracted for 1,000 random points distributed throughout the aforementioned directional distribution ellipse, and correspondingly for our surveyed sample of 394 HCT tombs. Comparison of descriptive statistics of flow accumulation for HCTs and random points supports the conclusion that tombs are indeed associated with water-rich areas (Table 22.1). The Kolmogorov-Smirnov (K-S) statistic test of the null hypothesis that these two distributions are the same against the alternative hypothesis that flow accumulation for HCT locations tends to be larger than those of random points gave a Z-value = 8.616, with a one-tailed p-value = 0.000, which further supports the conclusion that HCTs tend to be preferentially distributed in or near water-rich areas.

22.5   Concluding Remarks

South Arabian tombs and monuments offer an informative material record of ancient land-use, social relations and affiliations. Death, burial, ritual, and ancestors – often physically marked by cairns, monuments, and rock art – provide mnemonic devices for oral tradition, which is in turn an important tool in constituting and reproducing social frameworks.

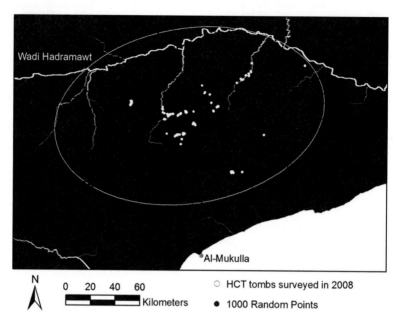

Fig. 22.5 Flow accumulation analysis map showing 500 m radius mean flow accumulation layer (background) with sample universe ellipse, random points and surveyed HCT tombs

Table 22.1 Flow accumulation descriptive statistics for random points versus HCT tombs

		Random points	HCTs
N		1,000	394
Mean		3,313	7,955
Minimum		1	2
Percentiles	10	3	12
	25	6	85
	50	18	1,523
	75	101	10,405
	90	752	25,158
Maximum		547,338	80,760

Our efforts to detect tombs over a substantial region have relied on high precision GPS field data that provides basic training sets for automated detection and spatial analyses. We expect that increasingly effective and streamlined implementations of automated detection methods will eventually offer tools for tomb mapping across large parts of southern Arabia particularly as high resolution imagery becomes less expensive and more widely available. Automated detection will likely also prove attractive in other regions, including in parts of Central Asia and South America, where morphologically comparable small circular stone tombs of an entirely different cultural and historical genealogy are similarly part of mortuary landscapes that speak to the palimpsest movements and affiliations of ancient life.

Both surveyed and detected datasets offer informative opportunities for spatial analyses readily implemented in GIS. In the present case our initial GIS analysis focused on spatial distribution of tombs in relation to water flow. Our results support previous assertions that HCT tombs are linked with water resources (Harrower 2008a), and though our survey was not designed to generate a random sample of tombs and is therefore subject to the vagaries of opportunistic survey, quantitative results do provide crucial support that moves such assertions beyond mere speculation. Future GIS analyses will explore tombs distribution in relation to viewsheds (cf. Bongers et al. 2012) and least-cost path travel routes, both of which are likely associated with landscape patterning of tombs and other monuments.

Acknowledgements This chapter briefly reviews some of the methods and preliminary results generated by the multidisciplinary AHSD research team led by Joy McCorriston and supported by National Science Foundation Grant BCS#0624268. Many participants deserve thanks including Dorota Brzezinska, Jihye Park, Tara Steimer-Herbet, Catherine Heyne, Jennifer Everhart, Kimberly Williams, Abdalaziz Bin 'Aqil, Khalid Badhofary, Abdalkarim Barkany, and Ietha Al-Amary. We are also grateful to the Republic of Yemen General Organization for Antiquities and Museums (GOAM) inclluding Abdullah Ba Wazir, and to Canadian Nexen Petroleum including Gregor Mawhinney, Alan Brindley, Kevin Tracy, Rick Jensen, Dave Smith and Kevin Marlow for their support.

References

Bin'Aqil, A., & McCorriston, J. (2009). Prehistoric small scale monument types in Hadramawt (southern Arabia): Convergences in ethnography, linguistics and archaeology. *Antiquity, 83,* 602–618.

Bongers, J., Arkush, E., & Harrower, M. (2012). Landscapes of death: GIS-based analyses of chullpas in the western Lake Titicaca basin. *Journal of Archaeological Science, 39,* 1687–1693.

De Cardi, B., Collier, S., & Doe, B. (1976). Excavations and survey in Oman, 1974–1975. *Journal of Oman Studies, 2,* 101–187.

De Maigret, A. (1996). New evidence from the Yemenite "turret graves" for the problem of the emergence of the south Arabian states. In J. Reade (Ed.), *The Indian Ocean in antiquity* (pp. 321–337). London: Kegan Paul.

Deadman, W. M. (2012). Defining the Early Bronze Age landscape: A remote sensing-based analysis of Hafit tomb distribution in Wadi Andam, Sultanate of Oman. *Arabian Archaeology and Epigraphy, 23,* 26–34.

Digital Globe Inc. (2012) Digital Globe Core Imagery Product Guide. Longmont, CO

ESRI (Environmental Systems Research Institute). (2011). *ArcGIS desktop: Release 10.* Redlands: Environmental Systems Research Institute.

Frifelt, K. (1975). On prehistoric settlement and chronology of the Oman peninsula. *East and West, 25,* 359–424.

Giraud, J. (2010). Early Bronze Age graves and graveyards in the eastern Ja'alan (Sultanate of Oman): An assessment of the social rules working in the evolution of a funerary landscape. In L. Weeks (Ed.), *Death and Burial in ancient Arabia and beyond: Multidisciplinary perspectives* (pp. 71–84). Oxford: Archaeopress.

Harrower, M. (2008a). Hydrology, ideology, and the origins of irrigation in ancient Southwest Arabia (Yemen). *Current Anthropology, 49*(3), 497–510.

Harrower, M. (2008b). Mapping and dating incipient irrigation in Wadi Sana, Hadramawt (Yemen). *Proceedings of the Seminar for Arabian Studies, 38*, 187–202.

Harrower, M. (2010). Geographic Information Systems (GIS) hydrological modeling in archaeology: An example from the origins of irrigation in Yemen. *Journal of Archaeological Science, 37*(7), 1447–1452.

Harrower, M., McCorriston, J., & Oches, E. A. (2002). Mapping the roots of agriculture in Southern Arabia: The application of satellite remote sensing, global positioning system and geographic information system technologies. *Archaeological Prospection, 9*, 35–42.

Harrower, M., Oches, E. A., & McCorriston, J. (2012). Hydro-geospatial analysis of ancient pastoral/agro-pastoral landscapes along Wadi Sana (Yemen). *Journal of Arid Environments, 86*, 131–138.

Harrower, M., Senn, M., & McCorriston, J. (in press). Tombs, triliths and oases: The Ancient Human Social Dynamcis Project (AHSD) archaeological survey 2009–2010. *Journal of Oman Studies*.

Maidment, D. R. (Ed.). (2002). *ArcHydro: GIS for water resources*. Redlands: ESRI Press.

McCorriston, J., et al. (2002). Holocene paleoecology and prehistory in Highland Southern Arabia. *Paleorient, 28*(1), 61–88.

McCorriston, J., et al. (2005). Foraging economies and population in the Middle Holocene highlands of southern Yemen. *Proceedings of the Seminar for Arabian Studies, 35*, 143–154.

McCorriston, J., et al. (2011). Gazetteer of small-scale monuments in prehistoric Hadramawt, Yemen: A radiocarbon chronology from RASA-AHSD Project research 1996–2008. *Arabian Archaeology and Epigraphy, 22*, 1–22.

McCorriston, J., et al. (in press) Monuments and landscape of mobile pastoralists: The Dhofar Monument Survey 2009–2011. *Journal of Oman Studies*.

Schuetter, J. M. (2010). *Cairn detection in southern Arabia using a supervised automatic detection algorithm and multiple sample data spectroscopic clustering*, Ph. D. dissertation, The Ohio State University.

Schuetter, J. M., et al. (in review). Autodetection of small-scale tombs in an Arabian tribal landscape in high resolution satellite images. *International Journal of Remote Sensing*.

Steimer-Herbet, T. (1999). Jabal Ruwaik: Megaliths in Yemen. *Proceedings of the Seminar for Arabian Studies, 29*, 179–182.

Steimer-Herbet, T. (2001). Results of the excavation in Jabal Jidran (February 1999). *Proceedings of the Seminar for Arabian Studies, 31*, 221–226.

Steimer-Herbet, T. (2004). *Classification des sépultures à superstructure lithique dans le Levant et l'Arabie occidentale*. Oxford: Archaeopress.

Yule, P., & Weisgerber, G. (1998). Prehsitoric tower tombs at Shir/Jaylah, Sultanate of Oman. *Beiträge zur Allegmeinen und Vergleichenden Archäologie, Band, 18*, 183–241.

Index

Made in the USA
San Bernardino, CA
17 January 2017